Eternal Stories from the Upanishads

Eternal Stories from the Upanishads

THOMAS EGENES
KUMUDA REDDY

Winsome Books India

First Edition: 2002
Reprint: 2004

The Sanskrit quotations from the Upanishads at the end of most stories are from translations by Maharishi Mahesh Yogi.

Published in 2009 by
Winsome Books India
209, F-17, Harsha Complex, Subhash Chowk
Laxmi Nagar, Delhi-110092
Email: winsomebooks@rediffmail.com
Website: www.winsomebooks.com

With deepest gratitude to
His Holiness Maharishi Mahesh Yogi,
who has brought to light the full meaning
of the Upanishads and the practical
technologies of consciousness to enliven
this knowledge in everyone, so that
all may rise to enjoy perfection in life.

Acknowledgements

We would like to thank Dr. Vernon Katz, Drs. Susan and Michael Dillbeck, Roxie J. Teague, Mary Zeilbeck, Janet Adelson, Caree Connet, Cheryl Johnson, and Michael Sternfeld for their valuable insights and additions; Susan Shatkin, Patricia Oates, and Dr. James Karpen for editing; and Martha Bright, Claudia Petrick, and Burton Milward for proofing, and Peter Freund for technical assistance. We would also like to thank Linda Egenes, who wrote the first two stories and edited all the stories.

CONTENTS

LIST OF COLOUR PLATES

Introduction

The Upanishads are a precious aspect of the Vedic Literature of India, the land of the Veda. It is the good fortune of the world at this time that Maharishi Mahesh Yogi has gathered the scattered, thousands-of-years-old Vedic Literature into a complete science of consciousness for its full theoretical and practical value.

Maharishi has brought to light a profound understanding of Veda and the Vedic Literature—that they comprise the basic structure of Natural Law at the basis of the universe. Veda and the Vedic Literature are the unseen, fundamental impulses of intelligence at the basis of the orderly evolution of the ever-expanding universe—which includes the orderly evolution of individual life.

With this understanding, we appreciate that the full meaning of the Upanishads is not found in books. Rather, the Upanishads are structures of our own intelligence, our own consciousness, our Self, and can be directly experienced in the simplest state of our own awareness. While reading stories from the Upanishads, it is important to remember that they are about the qualities of pure consciousness. Even though the stories describe the comings and goings of people and events, at a more subtle level of understanding, these stories describe the dynamics of consciousness found within everyone.

The connection between the Vedic Literature and our own consciousness can be clearly seen in the recent discovery of Professor Tony Nader, M.D., Ph.D.* Under Maharishi's guidance, Dr. Nader has found a precise correspondence between the different aspects of Veda and Vedic Literature and the structures and functions of human physiology. This discovery shows that every one of us is Veda. Every one of us has the total intelligence of Natural Law and its infinite organizing power within our own mind and body.

The beautiful, evolutionary qualities of consciousness expressed by the Upanishads and all other aspects of the Vedic Literature—qualities such as unifying, harmonizing, enlightening, transcending, and blossoming of Totality—are enlivened in the individual through the practical technologies of consciousness of Maharishi Vedic Science℠ and the Maharishi Transcendental Meditation℠ and TM-Sidhi® programs. The result of practicing these technologies is that our thinking and behavior become more creative, life-supporting, and free from mistakes—more in harmony with Natural Law, and therefore more and more successful and fulfilling.

This is the daily experience of millions of people throughout the world who practice the Transcendental Meditation® and TM-Sidhi programs—that health, happiness, mental potential, and harmonious relationships grow naturally and spontaneously. Over 600 scientific research studies conducted at more than 200 universities and research institutions in 30 countries have confirmed the benefits of the Transcendental Meditation and TM-Sidhi programs for all aspects of life—mind, body, behavior, and society. Those who practice these technologies will find, as they read and re-read these stories from the Upanishads, deeper meaning and connection with this knowledge, which describes the reality of their own intelligence, their own Self.

The Upanishads especially focus on the ultimate reality of life; they express the full glory of the Self, Ātmā, by gaining which nothing else is left to be gained. The Upanishads bring out that the true nature of the Self is wholeness, the totality of Natural Law, *Brahman.* From this level of experience, everyone and everything is near and dear to us as our own Self; one flows in universal love, nourishing everyone and everything.

Traditionally, the Upanishads were passed down from teacher to student. "Upa-ni-shad" literally means "to sit down near." Maharishi explains this as "everything sits down near the Veda." In other words, when we know the essence of everything to be Veda, then we have gained the fruit of all knowledge.

The Upanishads contain beautiful and exhilarating phrases such as "Thou art That" (*Tat tvam asi*), "I am Totality" (*Ahaṃ Brahmāsmi*), and "All this is Brahman—Totality" (*Sarvaṃ khalv idaṃ Brahma*). These phrases are nothing less than descriptions of the supreme awakening of consciousness to its own true nature. They are known as "great sayings" (*mahāvākya*) because they describe the essential teaching of the Upanishads in compact expressions. Maharishi describes these sayings as the final strokes of knowledge from the teacher, which fully enlighten the student who is ready to receive them; then Wholeness dawns in the awareness. In reading the stories from the Upanishads, we are thus reminded of the flow of our life towards its supreme goal.

Maharishi explains that the Upanishads, like all other aspects of Veda and the Vedic Literature, were cognized by the great enlightened Vedic Ṛishis, or seers; the profound truths dawned spontaneously in the silent depths of their own pure consciousness. Their cognitions are expressed in the language of nature, Sanskrit. According to the Muktikā Upanishad (1.30–9), there are 108 Upanishads, with ten principal Upanishads (*Īsha, Kena, Katha, Prashna, Muṇḍaka, Māṇḍūkya, Taittirīya, Aitareya, Chhāndogya,* and *Bṛihadāraṇyaka*). In this book, the name of the Upanishad is written under each story's title. Sanskrit words and phrases that appear in each story are listed at the end of the book, along with their pronunciation and meaning.

*Maharishi founded the Global Country of World Peace on October 7, 2000, and on October 12th he crowned Professor Tony Nader as the first sovereign ruler of the Global Country of World Peace with the title 'His Majesty Raja Nader Raam.'

अयमात्मा ब्रह्म

Ayam Ātmā Brahma
This Self is Brahman.
Māṇḍūkya Upanishad, 2

Chapter One

Satyakāma—the Seeker of Truth

From the Chhāndogya Upanishad

Long ago, in a small hut in the dense forests of India, lived a boy and his mother. The boy's mother named him Satyakāma, which means "seeker of truth."

More than anything, Satyakāma wanted to live the life of a student, meditating and studying about *Brahman* in the dwelling of his teacher. But to become a student, he had to know his father's family name, because in those times teachers only accepted students from certain families.

So Satyakāma went to his mother, Jabālā, and said, "Mother, I want to live the life of a student of sacred knowledge." Jabālā was pleased with her son's desire to study Brahman.

"Dear Mother, of what family am I?" Satyakāma asked.

"I do not know your family name, my precious son," his mother said.

"Then what shall I tell my teacher, dear Mother?" asked Satyakāma earnestly.

Jabālā led a pure life and knew the power of truth. "Tell him just what I have told you, my beloved son," she said.

With his mother's blessings, Satyakāma left his boyhood home. He walked through thick forests where the light of the sun never touched the ground. He saw foaming streams splashing on rocks. He passed by lakes as still and glassy as ice.

Soon he came to the home of the great teacher Gautama, who lived in his *āshram*, his Vedic school, by the edge of the forest.

Satyakāma bowed to the teacher in respect. "Please, honored Sir," he asked, "will you accept me as your disciple? With your blessings, I wish to become a knower of Brahman."

Gautama thought the boy looked healthy and bright. But to accept him as a student, Gautama had to know the boy's family background. And so he kindly asked, "Of what family are you, my boy?"

"My mother said to tell you that her name is Jabālā and my name is Satyakāma—and I know nothing more about my family," Satyakāma explained without fear. "So I am Satyakāma Jābāla."

Gautama was pleased that the boy's mother had taught her son to tell the truth. "Only one from the best of families could give this explanation so sincerely," he said. "I will gladly accept you as my student. Bring the firewood inside, my dear."

Satyakāma's heart felt warm with happiness. At last he would be able to study the knowledge of Brahman.

The next day Gautama said, "I will now begin teaching you the knowledge of Brahman, which is called supreme knowledge (*Brahma Vidyā*). The first step is to know your Self." And so Gautama initiated Satyakāma in meditation to settle his mind and heart. With a quiet mind, Satyakāma experienced his own inner Self, which was like a vast ocean of silence.

After teaching Satyakāma to meditate, Gautama did something that was very unusual. He took Satyakāma to the pasture where hundreds of cows were grazing. To Satyakāma's surprise, Gautama separated out four hundred thin, weak cows.

"Take these cows to another part of the forest and live, my dear boy," he said. "Tend them carefully. You may return when they have multiplied to a thousand!"

Without any doubts in his heart, the obedient Satyakāma drove his four hundred cows to a lush meadow on the other side of the forest.

At first Satyakāma felt lonely, since he was all by himself in the forest. But he sang to the cows and they mooed back to him as he slept. Satyakāma began to enjoy his life in the forest. His cows ate nourishing green grass and drank pure water from a spring-fed pond. Satyakāma watched his cows grow plump and happy.

Satyakāma stayed many years with the cows, living a peaceful life in the warm grassy meadows and cool forest. His days began and ended with meditation. As his mind became more and more quiet, he was able to comprehend the profound knowledge his teacher had given him.

He carefully tended the cows, always finding rich pastures for their grazing. As Satyakāma grew older, the herd of cows began to multiply. However, he was so contented with his life that he noticed neither the passage of time nor the increasing size of his herd.

Nevertheless, a profound change was taking place in Satyakāma. In his peaceful life in the forest, he was coming to know the Self. His mind became serene, his heart filled with love, and his face glowed with light.

Satyakāma never felt alone. He became friends with the proud peacocks, the rippling streams, and the swaying trees. He even became friends with the sun and the moon. Every living creature became part of his family. He remembered the phrase his mother had taught him, "The world is my family" (*Vasudhaiva kutumbakam*).

At night, as the cows slept, Satyakāma gazed at the infinite span of stars scattered across the sky like a thousand sugar crystals. He felt as if all nature was speaking to him. In the bright morning, the dew bathed his feet. White gardenias called to him with their sweet scents. Wispy clouds and distant rainbows delighted his imagination. Cool rain splashed his skin.

"All this beauty is a part of Brahman," he thought. "Everything that grows and decays is part of the great totality." He felt that he, too, was a part of the eternal cycle of life.

One day, the head of the cows, a wise bull, spoke to him. "Satyakāma!"

"Yes, honored sir," answered Satyakāma, who respected all living things.

"We are now one thousand cows," said the bull. "Please take us to your teacher's hermitage. And I will teach you the nature of Brahman, which has many aspects."

"Yes, please tell me," said Satyakāma.

"Brahman shines from the east and the west," the bull told him, "and from the north and south. This is because Brahman is everywhere. It is universal. This is one quarter of Brahman."

Then the bull said, "The fire, Agni, will teach you more about Brahman."

Satyakāma began to drive the cows back to his teacher's āshram. When evening came, he put a rope around a large area to protect the cows. Then Satyakāma lit a fire. He sat down on the west side of the fire, facing east. He gazed at the dark sky, filled with his friends, the stars. After some time, the fire, Agni, spoke to him about the nature of Brahman.

"Brahman is the earth and the atmosphere," said Agni. "It is the sky and the ocean. This is because Brahman is endless. It is without beginning or end. This is one quarter of Brahman."

Then Agni added, "A swan will tell you more about Brahman."

The next evening, after traveling with his cows, Satyakāma again lit a fire by the side of a river. He sat down, facing east, and saw a great white swan gliding down the river towards him. The swan began to teach Satyakāma about the nature of Brahman.

"Brahman is fire," the swan explained, "and Brahman is the sun. Brahman is the moon, and Brahman is lightning. This quarter of Brahman is light. Brahman is the light of life."

Then the swan said, "A bird will tell you more about Brahman."

The next evening, after settling his cows in a safe place beside a hill, Satyakāma again lit a fire. He sat down on dry, soft grass, facing east. This time a purple sunbird flew down from the limb of a tree. Like silk woven with gold, its wings caught the brightness of the fire.

The sunbird sang, "Brahman is the breath, and Brahman is the eye. Brahman is the ear and also the mind. This quarter of Brahman is the seat, the resting place. Just as the eye is the seat of what is seen, and the mind is the seat of what is thought, so Brahman is the seat of everything. Everything rests upon Brahman."

Finally Satyakāma arrived at his teacher's dwelling. His teacher noticed how Satyakāma's face was shining, and he said to him, "I see that you have found Brahman. For it is said that the knower of Brahman has settled senses, a smiling face, freedom from worry, and has found the purpose of life."

But even with these words of praise, Satyakāma spoke humbly, "Please, honored Sir, teach me about the nature of Brahman." For Satyakāma wanted to learn from his teacher about the true nature of Brahman.

"You have heard that east and west are Brahman, that earth and sky are Brahman, that the sun and moon are

Brahman, and that the eye and ear are Brahman," his teacher replied. "Like waves stirring within the ocean, all these are a part of Brahman. This is because Brahman is everywhere. Brahman is everything (*Brahmaivedaṃ sarvam*). It is endless. It is the light of life. Everything finds its rest in Brahman.

"And Brahman is realized by knowing the Self, your true nature. Then you realize that you are everywhere—you are endless, and you are radiant. This is the supreme knowledge, Brahma Vidyā. Yes, this is the supreme knowledge, Brahma Vidyā."

And that is how Satyakāma came to know Brahman, and grew up himself to become a great teacher of Brahman.

सर्वं खल्विदं ब्रह्म

Sarvaṃ khalv idaṃ Brahma

All this is Brahman—Totality.

Chhāndogya Upanishad, 3.14.1

CHAPTER TWO

The Story of Shvetaketu

From the Chhāndogya Upanishad

Many years ago in a village in India there lived a boy named Shvetaketu. Shvetaketu was the son of a great and wise man, Uddālaka.

When Shvetaketu turned twelve, his father said to him, "My dear son, follow the tradition of our family and become a man of knowledge." And so, as was the custom then, Shvetaketu went to the home of his teacher to study.

After many years, he finished his studies and returned to his father's house. His father noticed that something was different about his son Shvetaketu. Shvetaketu thought he knew everything there was to know.

Shvetaketu's father saw his son's pride. "My dear son," he asked, "I wonder if you have learned the wisdom of the Veda. Can you hear what cannot be heard by the ear? Can you see what cannot be seen by the eyes? Do you know what cannot be known by the mind?"

"What do you mean, dear Father?" asked Shvetaketu in surprise.

"Do you know that by knowing which everything else becomes known?" asked his father.

"What is this teaching, Father?" asked Shvetaketu.

"My son, when you know one lump of clay, you know all that is made of clay.

"When you know one nugget of gold, you know all that is made of gold.

"When you know one pair of iron tongs, you know all that is made of iron."

"I have not yet learned this teaching," said Shvetaketu humbly. "Please, Father, teach me."

"As you wish, my dear son," said his father.

"In the beginning there was an unbounded ocean of consciousness, one without a second (*Ekam evādvitīyam*). The ocean of Being thought to itself, 'I am one—may I be many,' and created light. Light thought to itself, 'I am one—may I be many,' and created water. Water thought to itself, 'I am one—may I be many,' and created matter.

"One unbounded ocean of consciousness became light, water, and matter. And the three became many. In this way the whole universe was created as an unbounded ocean of consciousness ever unfolding within itself.

"That infinite source of the whole universe, the Self of all that is, the ocean of pure consciousness, that essence of all things—that is truth. That is the unbounded Self. Of that you are created. That thou art (*Tat tvam asi*), Shvetaketu."

"Please, honored Sir, tell me more of this teaching," said Shvetaketu.

"As you wish, my dear son," said his father. "Come with me to the orchard."

When they arrived at the orchard, he said, "See the bees collecting nectar? Once the nectar is gathered, it does not say, 'I am the essence of the apple blossom,' or 'I am the essence of the orange blossom.' No, the nectar joins with itself, and is called honey.

"In the same way, Shvetaketu, when people contact the ocean of pure consciousness, they become one with it and do not remember their individual natures. Yet when they are active, they again become a teacher, a farmer, or a goldsmith.

"That subtle essence of the whole world, the Self of all that is, the ocean of pure consciousness—that is truth. That is the eternal Self. Of that you are created. That thou art (*Tat tvam asi*), Shvetaketu."

"Please, honored Sir, teach me more," Shvetaketu said.

"As you wish, my dear. Come with me to the river."

When they arrived at the river, they stood at its banks and watched the water rushing by. His father said, "As the rivers flow to the east and merge with the sea, they become the sea itself. Once they are the sea they do not think, 'I am the river Gangā,' or 'I am the river Kshiprā.' They know, 'I am the sea.'

"In the same way, dear Shvetaketu, even though all creatures emerge from the ocean of consciousness, they do not know that. Whatever they are—whether tiger, lion, or wolf—in the end they return to the ocean of consciousness.

"That subtle essence of the whole world, the Self of all that is, the ocean of pure consciousness—that is truth. That is the Self, which unifies. Of that you are created. That thou art (*Tat tvam asi*), Shvetaketu."

"Please, dear Father, teach me more," Shvetaketu said.

"As you wish, my son," said his father. "Bring me a fruit from the banyan tree."

Shvetaketu went outside and picked a fig from the long branches of the banyan tree. "Here it is, Father," said Shvetaketu.

"Break it open," said his father. "Tell me, what do you see inside?"

"I see many seeds."

"Break one seed open," said his father. "Tell me what you see."

"I see nothing at all," said Shvetaketu.

"My son, that 'nothing' is the subtle essence of all living things, which appears as nothing because you cannot perceive it. But from that nothing this great and ancient tree has grown.

"That infinite source of the whole universe, the Self of all that is, the ocean of pure consciousness—that is truth. That is the unmanifest Self. Of that you are created. That thou art (*Tat tvam asi*), Shvetaketu."

"Please, honored Sir, teach me more," said Shvetaketu.

"As you wish, my dear son," said his father. "Fill this glass with water and add some salt. Then bring it to me in the morning."

Shvetaketu did this. The next morning his father said, "Bring me the salt you poured into the glass."

Shvetaketu returned with the glass of water and said, "The salt has disappeared, Father."

"Please take a sip from the top of the glass," his father said. "How does it taste?"

"Salty."

"Now pour out some and take a sip from the middle," said his father. "How does it taste?"

"It tastes salty."

"Now pour out some and take a sip from the bottom. Tell me, how does it taste?"

"Salty," said Shvetaketu.

"Even though you couldn't see it, the salt was found in every drop of water. In the same way, pure consciousness is found in all beings. That subtle essence of the whole world, the Self of all that is, the ocean of pure consciousness—that is truth. That is the all-pervading Self. Of that you are created. That thou art (*Tat tvam asi*), Shvetaketu."

"Please, honored Sir, teach me more."

"As you wish, my dear son," said his father. "Think of a man left blindfolded in a desert. He wanders around, not knowing where to go. But if someone removes the blindfold and points out the right direction, he finds his way until finally he reaches home. In the same way, if a teacher points the way to Self-knowledge, then you enjoy the path to enlightenment—from the very first step you grow in intelligence, happiness, and success in life.

"That subtle essence of the whole world, the Self of all that is, the ocean of pure consciousness—that is truth. That is the Self, which removes the darkness of ignorance. Of that you are created. That thou art (*Tat tvam asi*), Shvetaketu.

"When you have known this subtle essence of life, then you have seen the unseen and known the unknown. You have known that by knowing which everything else becomes known."

And then Shvetaketu understood the true teaching of the Veda. Even when he later became a famous teacher in the court of King Janaka, Shvetaketu always remained humble, once he had realized the Self, the ocean of pure consciousness.

अथ यदतः परो दिवो
ज्योतिर्दीप्यते विश्वतः पृष्ठेषु
सर्वतः पृष्ठेष्वनुत्तमेषूत्तमेषु लोकेषु
इदं वाव तद्यदिदमस्मिन्नन्तः पुरुषे ज्योतिः

Atha yad ataḥ paro divo
jyotir dīpyate vishvataḥ pṛishtheshu
sarvataḥ pṛishtheshv anuttameshūttameshu lokeshu
idaṃ vāva tad yad idam asminn antaḥ purushe jyotiḥ

There is a light which shines beyond the world,
beyond everything, beyond all, beyond the highest heaven.
This is the light which shines within your heart.

Chhāndogya Upanishad, 3.13.7

Finer than the finest, greater than the greates

CHAPTER THREE

Nachiketas Gains Immortality

From the Katha Upanishad

Once there was a young boy named Nachiketas, who had many good qualities. He was intelligent, happy, and patient. His wise and peaceful heart shone like the full moon on a clear night.

One day his father, Vājashravas, began a certain *yagya* (a performance that creates balance in nature). Nachiketas felt his heart swell in happiness as the *pandits* (Vedic scholars) offered milk, ghee, and rice. The fire burned brighter and brighter as they chanted the Vedic hymns in a gentle rhythm that was as old as time itself.

After the yagya, Nachiketas watched his father present gifts to the pandits—cotton shawls, mangoes from their garden, and many cows.

As he looked more closely at the cows his father was giving away, the observant Nachiketas suddenly felt worried.

"Unfortunately," he thought, "these cows are weak and thin, and have no more milk. What will happen to the person who gives such poor old cows? Why did my father select his worst cows? Shouldn't he give the plump ones with lots of milk in their udders?"

So the respectful Nachiketas thought to offer himself as a gift. He said to his father, "To whom will you give me?"

Vājashravas realized what his son was saying, but ignored him.

"Father, to whom will you give me?" Nachiketas asked again.

Still his father did not answer. By this time his head was swelling with anger, for he knew his son was pointing out his own lack of generosity.

"To whom will you give me, dear Father?" Nachiketas persisted.

Finally his father burst out, "Unto Yama I give you!"

Nachiketas knew in his heart that his father loved him dearly and that he had only spoken these words in anger.

"My dear son, I do not really wish you to go to Yama," his father said. But though Vājashravas tried to take them back, the obedient Nachiketas felt he should honor his father's words.

So Nachiketas went south to the world of Yama, the administrator of death and immortality. There Nachiketas found Yama's house, surrounded by an iron fence.

Yama was not at home. So Nachiketas sat on the doorstep and waited. He waited and waited, and still no sign of Yama. Nachiketas, who was very patient, waited there for three days and three nights. He had no food to eat or water to drink, and no bed to sleep in.

Finally, Nachiketas saw Yama approaching, riding on a water buffalo. On his head Yama wore a golden crown, showing everyone that he rules his own world. He wore red robes and his skin was gray. His eyes glittered like cool gems above his long black mustache.

Yama looked fierce even to the brave Nachiketas. But when Yama started to speak, he smiled and greeted the boy kindly.

"What is such a bright, healthy young man doing on my doorstep?" he asked in surprise.

Nachiketas explained all about his father and the weak cows.

"And you have waited for three days and three nights without eating a morsel or drinking a drop or sleeping a wink!" exclaimed Yama.

Yama was very worried. He knew that he should have treated his young guest better. He remembered the saying, *Atithi devo bhava*—"Honor the guest as God."

Yama wanted to make up for his lack of hospitality. So he said, "Nachiketas, my honored guest, you may have three boons. Ask for three wishes, whatever you want, and be gone."

"For my first wish," Nachiketas answered, "I would like my father's anger to disappear, and I would like him not to worry about me. I would like him to be happy to see me, once I am set free by you."

"As you wish, Nachiketas," Yama replied. "Your father will sleep peacefully through the nights, and will greet you with joy. Choose your second boon."

Nachiketas thought for a while. Then he said, "O Yama, please describe the yagya which will help me reach heaven."

Yama then taught Nachiketas each detail for performing the yagya. Yama taught him the size of the bricks, how to build the altar, and the Vedic hymns to recite. Yama taught Nachiketas how to sprinkle the water, what to offer to the fire, and how to make the offerings while saying *svāhā*, "Hail!"

The clear-minded Nachiketas repeated everything exactly as he had been taught. Yama was pleased with him and said, "From now on, this yagya will be known as the Nāchiketa Fire, in honor of your devotion to knowledge."

Then Yama said, "Choose your third and final boon, O Nachiketas."

"O Yama, teach me the nature of immortality," Nachiketas said. "If I gain immortality, will I still be myself after I die? Or will I become one with the ocean and sky and earth?"

"O Nachiketas, even the *Devas*, even the powers of nature, have doubt on this point," Yama replied in a worried voice. "It is not easy for someone so young to understand, so subtle is this truth. Choose another boon. Release me from this obligation. Do not press me!"

Nachiketas remained steadfast. "No other boon is the equal of this. Please, Sir, instruct me on this point."

"Choose children and grandchildren," Yama begged. "Choose cattle, horses, elephants, and carts filled with gold. Choose vast tracts of land, choose long life for yourself, choose fame, choose swift chariots and dance and song. But do not ask for this, O Nachiketas!"

The wise Nachiketas shook his head. He wanted more than these material things. "Please, Sir, teach me about the mystery beyond human life. Tell me about the nature of immortality. I ask for no other boon."

"You are very intelligent, Nachiketas," Yama finally said, "to choose knowledge instead of worldly things. I will grant you this boon. Here is the secret of immortality."

Then Yama began his instruction: "There are two paths in life—the path of pleasure and the path of spirituality. The first is ignorance and the second is wisdom. They are very different and lead to very different ends.

"The person who seeks pleasure is left with nothing in the end. The one who seeks spirituality obtains Brahman—the totality, the fulfillment of all desires. He rejoices, having attained the source of joy.

"Through a tranquil mind one realizes the Self, which is set in the heart of every creature. This is Brahman, finer than the finest and greater than the greatest (*Aṇoraṇīyān mahato mahīyān*)."

"Tell me more about the Self, O Yama, so that I may understand," Nachiketas said.

You have seen the chariots of the king?" asked Yama.

"Yes," replied Nachiketas.

"Imagine that your body is the chariot, your intellect is the charioteer, and your mind is the reins," Yama said. "Your senses are the horses, and the objects you see in front of you are the roads. The Self is like the lord of the chariot, who rides in back.

"When your intellect (the charioteer) chooses rightly, then the reins stay taut and the horses act like good horses. You can reach your goal easily in such a chariot.

"But when your intellect chooses wrongly, then the reins come loose, and the horses act like wild horses. You will never reach your destination in such a chariot.

"If a person does not know the Self, then his mind is restless, like uncontrolled reins. The horses dash this way and that, dragging the whole chariot in every direction. Such a person never reaches the goal of life.

"But those who know the Self have even minds. They reach their home. They peacefully come to the end of their journey—Brahman."

"Please, Sir, tell me more," Nachiketas said.

"Beyond the objects of perception are the senses," Yama said. "Beyond the senses is the mind, which thinks. Beyond the mind is the intellect, which decides. Beyond the intellect is *Ātmā,* the Self. The Self is without sound, without touch, and without form.

"The Self is eternal—without beginning or end. By knowing the Self, you know the secret of immortality.

"When you seek life eternal, you must turn your attention inward. There you will find the Self. The innermost Self resides in the center of the heart like a flame without smoke. It is the same today and will be the same tomorrow. It grants all desires.

"The Self is Brahman. It is the immortal. You will know the Self when your senses are still, your mind is at peace, and your heart is pure."

Nachiketas was very grateful to Yama for teaching him. He said, "Thank you, Sir, for this knowledge. Please tell me, what should I do first?"

"Start by knowing that the Self exists," Yama said. "Then seek to learn more about it by experiencing it directly through a settled mind. When thought has ceased, and even the intellect does not stir, this is the highest state. This is *Yoga*, the state of union. This is how you will keep your heart pure and your mind peaceful.

"O Nachiketas, come to know your inner Self, which dwells in your heart. It should be firmly drawn out, as one draws out the air from the center of a reed. This is the pure. This is the immortal. Yes, this is the pure. This is the immortal."

And so the young but wise Nachiketas received from Yama the teaching of immortality and the way to achieve it. After returning to his father, he was freed from all impurities and obtained immortality. He attained Brahman. And so may anyone who truly knows the Self.

अणोरणीयान्महतोमहीयान्

Aṇoraṇīyān mahatomahīyān

Finer than the finest, greater than the greatest.

Katha Upanishad, 1.2.20

Chapter Four

Raikva the Cart Driver

From the Chhāndogya Upanishad

Long ago there was a great king named Jānashruti. His name meant "celebrated among the people." It was a fitting name, because he gave abundantly to all in his kingdom. King Jānashruti invited his subjects to lavish feasts, where he filled long banquet tables with delectable curries, baskets of sweets, and cool drinks.

In each village the king built fountains and lakes surrounded by beautiful gardens. He built smooth roads with rest houses where people traveling on a long journey could eat and spend the night.

"Everyone enjoys my food!" he thought. King Jānashruti liked giving to the people, for he felt that he received many blessings from his generosity.

One evening King Jānashruti was enjoying the cool night air on the marble verandah of his palace. Palm trees and sweet-smelling orchids surrounded the royal silk couch of the wise Jānashruti. A large jade statue of Ganesh, the remover of obstacles, brought good fortune to the famous king.

King Jānashruti rested on his couch and sipped warm milk and ghee (clarified butter) while his chief minister reviewed the day's reports from around the kingdom. The just king prided himself in governing all with fairness.

"We should like to have some fresh mangoes!" he said. Of course, when King Jānashruti made a request, everyone responded quickly, because he was the king, and also because they loved their kind ruler.

As the king relaxed on the verandah, the sun slipped below the horizon, and pink and orange clouds floated in the sky. In the distance, two geese flew gracefully toward the royal palace. The king watched the two geese glide nearer and nearer. As they came closer, the king could not help overhearing them talk to each other.

"Hey ho! Are you blind?" the first goose called to the other. "You must be careful not to fly across this stretch of sky. It is radiant with the splendor of the great-hearted King Jānashruti. Avoid flying through it, or you'll burn your wings!"

King Jānashruti listened contentedly from the marble verandah. He was glad to hear that even the birds knew of his generosity and wisdom.

Then he heard the second goose say, "Who are you talking about? You speak of his brilliance as if he were Raikva."

"Who is this Raikva?" asked the first goose.

"You know how in a game of dice, the winner, the one with the highest throw, takes all?" said the second goose. "Like that, the good deeds that everyone performs all over the kingdom flow to Raikva, because he is the wisest of all. He is the most generous. He gives the most to the people, and he receives the blessings of their good deeds." Then the geese flew away.

King Jānashruti couldn't believe his ears! How could someone else be wiser than himself? Wasn't he the most generous in the land? Who was Raikva, anyway? The king was so upset he barely slept that night in the royal chambers.

As on every other morning, the royal courtiers heralded the new day by singing the praises of the wise and generous king, "O Friend of All the World, we hail you! O Fountain of Glory, we. . . ."

Before they could finish, the king raised his hand and called out to the royal bards, "I cannot bear to hear your praises." Then he said quietly to his attendant, "On this morning I must know—who is Raikva?"

His faithful servant bowed and replied, "Your kind Majesty, I am very sorry, but I have never heard of Raikva. Who is this man?"

King Jānashruti told him about the conversation between the two geese. "Please search for Raikva," the king said. "I must meet him. Perhaps you will find him on the bank of a river, in a cool cave, or by the peaceful woods."

That day a delegation from the king began a great search for Raikva. They traveled to serene places where saffron-clad holy men with trident staffs practiced their meditations. They walked the banks of sacred rivers, lit the insides of dark caves, and searched the dense forests of the kingdom. But they could not find him among the *yogīs* and wise men.

Finally, the delegation gave up. On their way back to the royal palace, they stopped to rest in the crowded marketplace of a small village. The market was bursting with people and overflowing with food.

Mangoes, bananas, coconuts, dates, guavas, and papayas were piled high on wooden ox-carts in the cramped market. Mounds of *tīka* powder in red, pink, and blue formed colorful pyramids on the traders' mats.

"Come repair your sandals!" cried the shoe *wallah* (vendor).

"Only one *rupee* for a shave!" called out the barber from his stall.

"Two rupees to hem your *dhotī* (clothing)!" called the tailor from his shop.

Vendors fried *samosās* (spicy vegetables) and *pūrīs* (puffed bread) in vats of clarified butter. The pungent aroma of spices popping in ghee mingled with the sweet smell of hanging marigold and jasmine garlands.

Rickshaws darted in and out among the crowd. Goats and pigs wandered through the streets looking for food. The rooftops teemed with monkeys scampering and screeching, as if chattering about the high price of bananas. White cows napped safely in the middle of the street, flower blossoms adorning their painted horns.

Amidst the noise and din, the king's representatives sat drinking fresh juice from tender coconut shells. They were very tired, and the bustling marketplace made them feel more exhausted than ever.

As they sat, they noticed a man sitting in the shade under a bullock cart. His clothes were in rags, and his feet were dusty.

Even though he looked like a poor man, his large serene eyes and peaceful face created a feeling of silence amidst the clatter of the market. A radiant light shone all around him. He seemed to be floating in sweet, simple happiness.

"That man looks like a great *Ṛishi,* a great seer, from the Himālayas!" exclaimed one of the king's men.

He walked over to the man sitting under the cart and said, "Excuse me, Sir. What is your name?"

"I am Raikva," the man replied with a radiant smile.

The king's delegation forgot their tiredness. They felt jubilant and happy. They rushed back to the palace.

"Your Majesty, we have found him!" they proclaimed as they entered the royal court.

King Jānashruti was delighted, and decided that he himself would pay a visit to Raikva. He said, "I will meet this man tomorrow! I will bring him many expensive gifts!"

The king thought he would please Raikva with his generosity. Leading a grand parade of mighty elephants, strong soldiers, and beautiful chariots, the king left the palace and marched toward the small village where Raikva lived.

The villagers heralded King Jānashruti with the fanfare of trumpets and blaring of conches. "*Jai Rājā!* Hail the King! *Jai Rājā!*" the people chanted as the king passed by on his gold and red palanquin, resting on the shoulders of four attendants. Never before had the glorious King Jānashruti come to their village.

As the king entered the marketplace, he stepped out of his palanquin and stood up. The crowd cheered louder, but when he held up his hand for silence, not a whisper stirred among them.

As the villagers stared in disbelief, the king carried garlands of jasmine and marigolds to the bullock cart where the serene Raikva was sitting. Even the sleeping cows looked up to see what was happening.

"Greetings, O revered Raikva!" cried the king, for all to hear. "I wish to bestow upon you a gold necklace and a chariot. Choose six hundred cows from the royal herd. The royal court this day smiles upon you! Good fortune has come to you! Today these precious goods are yours!"

The king was sure that the poor Raikva would be overwhelmed with his precious gifts. But Raikva did not even look at them. He looked straight at the proud king. "I don't want your gifts," he said gently. "Please take them and leave."

Everyone gasped. No one had ever refused the king's gifts before. What would happen now?

The king said nothing. He simply returned to his palanquin and—with his mighty elephants, strong soldiers, and beautiful chariots—slowly rode back to the royal palace in silence.

Sad and dejected, the king closed himself in his royal chambers so that he could be alone to think. "What went wrong?" he wondered. "Why didn't the poor Raikva accept my gifts?"

The king remembered Raikva's serene eyes and peaceful face. He thought about his radiant light. He thought about Raikva's gentle smile. He thought and thought.

Suddenly the king knew what went wrong.

He realized that the gifts he offered—cows and gold—meant nothing to Raikva. Raikva owned inner happiness. Raikva owned spiritual knowledge. Raikva owned peace, simplicity, and bliss. The king saw that he himself owned only marble and gems.

"Raikva's spiritual happiness is much more valuable than my material gifts," thought the king. "All that I have means nothing to Raikva, but what he has means everything to me!"

Then the king knew he must return to Raikva.

Once again the king set out—with his mighty elephants, strong soldiers, and beautiful chariots—to visit the peaceful Raikva. This time the king would offer even more gold, cows, and chariots. But this time giving gifts was no longer important to the king. This time he went with humility.

Arriving in the village, the king bowed down to Raikva and said quietly, "Dear Sir, I humbly ask you to keep these gifts. What I wish with all my heart is for you to accept me as your student. Please Sir, teach me what true happiness is."

This time Raikva accepted the king's gifts. He saw that King Jānashruti had lost his pride, and was ready to learn. He saw that the king sincerely wanted knowledge.

And so Raikva taught the king the nature of Ātmā, the Self. "The Self does not need anything," he said. "It is self-satisfied and self-sufficient. He who knows the Self finds bliss. He finds satisfaction. His happiness lasts forever.

"If you know the real nature of the Self, you will never want anything. You will always be content."

Raikva then taught the king about giving. "Dear King," he said, "it is good that you like to help others, but do not give with pride. Remember that the things you give away are not yours. They are gifts from nature. Give them in all humility.

"And while it is good to give material possessions to the people, it is better to give spiritual welfare. Knowing the unbounded Self will make the people truly happy. Yes, this will make them happy."

Then Raikva taught the king how to practice meditation, which is the heart of the Vedic tradition of knowledge. The king was grateful to receive Raikva's gifts, and he returned to his palace.

The king practiced meditation each morning as the sun rose over his marble verandah and each evening as it set. He felt more peaceful and content each day.

One day he thought, "I want to give the people what I feel inside. I want to give them what Raikva gave me. I want to give them true happiness."

And so the king asked that all his subjects be taught meditation. After a short while, his people began to awaken early in the morning, feeling fresh and vital.

The people became more energetic and creative. Crime and sickness disappeared. They built beautiful homes and cities, and turned their kingdom into a garden. They celebrated festivals with singing and dancing. People easily followed their *Dharma*, their natural duty, and enjoyed satisfaction and contentment in their daily lives.

King Jānashruti became known as a great king, a *Mahārāja*, a spiritual king who had found fulfillment within himself and taught his subjects how to know true happiness.

यो वै भूमा तत्सुखं नाल्पे सुखमस्ति

Yo vai bhūmā tat sukhaṃ nālpe sukham asti

That which is unbounded is happy.

There is no happiness in the small.

Chhāndogya Upanishad, 7.23

CHAPTER FIVE

Indra Asks about the Self

From the Chhāndogya Upanishad

High in the starry heavens Prajāpati, the protector of life, was teaching about the nature of Ātmā, the Self.

"Ātmā is free from wrong, free from sorrow, free from hunger and thirst," he said. "Those who know Ātmā fulfill all their desires. They gain all worlds. If you want to know the truth, know the true nature of Ātmā, the Self."

Prajāpati's words were overheard by both the shining *Devas* (the positive powers of nature) and the *asuras* (the negative powers). Both the Devas and the asuras thought, "We must learn about Ātmā, the Self. It is through the Self that we will gain support from nature. By knowing the Self, we can fulfill all our desires."

So each group sent a representative to find out the nature of Ātmā from Prajāpati. The Devas sent Indra, their glorious leader, who wields the mighty thunderbolt. The asuras sent Virochana to represent them.

"Please, Sir, tell us the true nature of Ātmā, the Self," Indra and Virochana humbly asked Prajāpati.

"Bring a pan of water," said Prajāpati.

They brought a pan of water. "When you look into the water, what do you see?" Prajāpati asked.

"We see our faces and bodies, even our hair and nails," they answered.

"This is Ātmā," said Prajāpati. "This is the immortal. This is the spirit. This is the Self."

Upon hearing this, Indra and Virochana went home with peaceful hearts, thinking they had learned the nature of Ātmā. As they left, Prajāpati looked at them and thought, "They go away without having known the Self. They think that the body is the Self. Whoever thinks such a thing will not become enlightened."

After the meeting, Virochana went triumphantly back to the asuras and told them, "The body is Ātmā. If the body is satisfied, then we will obtain all our desires." They were very happy to have this teaching.

Indra, on the other hand, felt confused. He was driving his golden chariot across the sky, turning over in his mind the words Prajāpati had spoken. He did not even notice the jeweled palaces, golden rivers, or lotus lakes he passed over. Suddenly, he stopped his chariot in mid-air.

"Wait a minute!" he thought. "Something seems to be wrong. If the Self is the body, then the Self must die when the body dies. Therefore, the Self would not be immortal."

Indra decided to go back to Prajāpati, to find out more about Ātmā, the Self. Indra said, "Please, Sir, teach me more about the true nature of the Self."

Prajāpati was pleased with Indra's request. "O Maghavan, O Indra, he who plays happily in a dream, he is

Ātmā. He is the Self. He is pure happiness," Prajāpati said. With this answer, Indra left with a peaceful heart.

Indra started driving his golden chariot across the sky to bring the Devas this answer—that the person who dreams is the Self. Along the way, he started to think about Prajāpati's words. He did not even notice the blue-tinted mountains, still pools, or singing waterfalls he flew over. Suddenly, he stopped his chariot in mid-air.

"Wait a minute!" he thought. "The person who dreams can also experience unhappiness. He can have a bad dream. Surely this is not Ātmā. Surely this is not the Self, which is pure happiness."

Indra decided to return once again to Prajāpati, to find out more about Ātmā. "Please, Sir, teach me more about the true nature of the Self," he said.

Again Prajāpati was pleased with Indra's request. "O Maghavan," Prajāpati said, "when a person is asleep, still and tranquil, and is not dreaming, that is the Self. That is Ātmā. That is fulfillment." With this answer, Indra left with a peaceful heart.

Indra began driving back to the other Devas with this answer—that the person who sleeps is the Self. Along the way, he started to think about Prajāpati's words. He did not even notice the drowsy sun slipping behind the hills, the cool twilight wind weaving the air, or the soothing call of evening birds far below him. Suddenly he stopped his chariot.

"Wait a minute!" he thought. "The person who sleeps is not even aware that he is sleeping. How can he be fulfilled?"

Again Indra returned to Prajāpati to find out more about Ātmā. Folding his hands in respect, Indra patiently asked, "Please, Sir, teach me more about the true nature of Ātmā."

Prajāpati at last told Indra the true meaning of Ātmā. "O Maghavan, the body is mortal. But the Self is immortal—it lives forever. While associated with the body, the Self experiences pleasure and pain. When the Self is no longer associated with the body, it goes beyond pleasure and pain. The Self remains always the same. The Self is immortal.

"You have seen the sky?" asked Prajāpati.

"Yes, Revered Teacher," answered Indra.

"The sky is formless," continued Prajāpati, "but for a time it takes on the shape of clouds, lightning, or thunder. These appear and then vanish. The sky remains the same. Just like clouds, our bodies appear and then vanish. And like the sky, the Self remains the same. The Self is immortal.

"For a time, the Self becomes as if hidden by the golden world. It is said, 'The face of truth is hidden by a covering of gold' (*Hiraṇmayena pātreṇa satyasyāpihitaṃ mukham*). But then the Self shines through, because the Self is immortal.

"When the Self knows its own nature, it shines within itself. This is the highest light. This is Ātmā. When the Self knows its own nature, it laughs and plays and rejoices. The Self is pure happiness. The Self is immortal."

Then Prajāpati asked, "You have seen how the water buffalo is attached to the cart?"

"Yes, Sir," answered Indra.

"Just like that, the Self is joined to the body," continued Prajāpati. "The water buffalo is not always attached to the cart, however. Sometimes it plays in pools of water, or wanders in the fields eating grass. Like this, the Self becomes free. Established in the Self, one is eternally happy.

"When you look at the sky, O Indra, your eyes do the seeing, but the Self is the seer. This is why we say, 'I see the sky.' When you smell jasmine, your nose does the smelling, but the Self is the one who smells jasmine. This is why we say, 'I smell the jasmine.'

"When you speak, O Indra, your voice does the speaking, but the Self is the speaker. This is why we say, 'I am speaking.' When you listen to music, your ears do the hearing, but the Self is the listener. This is why we say, 'I hear the music.' It is the Self who is the enjoyer of bliss. It is the Self who is fulfilled."

Indra smiled and thanked Prajāpati. Then Indra returned to the Devas and told them what he had learned about the Self, Ātmā. The Devas came to know the true nature of the Self as immortal, as pure happiness, as fulfillment. Therefore the Devas achieved all their desires, and they obtained all worlds. Yes, they achieved all their desires, and obtained all worlds. And so can anyone who knows the Self.

तरति शोकमात्मवित्

Tarati shokam Ātmavit

Established in the Self,

one overcomes sorrows and suffering.

Chhāndogya Upanishad, 7.1.3

Established in the Self, one overcomes sorrows and suffering

CHAPTER SIX

The Devas and the Blade of Grass

From the Kena Upanishad

On the edge of a great forest long ago, the morning sun warmed the walls of a quiet āshram. Many young students lived there in tidy thatched cottages. Every day children sat under the spreading branches of an ancient banyan tree and recited the Vedic texts after their teacher.

One morning they gathered on the soft, cool grass under the tree. Sunlight flickered and danced through the lacy canopy of leaves, casting a kaleidoscope of designs on the students' yellow cotton *kurtās*, their traditional long shirts. They chattered happily to each other, waiting for their *āchārya*, their great teacher, to give the morning's lesson.

The āchārya appeared at the door of his cottage. His eyes sparkled like the warm and gentle morning sun. His sandals whispered in the grass as he walked. The students stood and folded their hands to greet him. After he sat down, the children gathered around and sat on the grass at his feet.

The teacher began his lesson with a blessing:

"May my arms and legs be strong.
May my speech be vital,

and also my breath, eyes, and ears.
May all my senses be healthy and vigorous.
Āpyāyantu mamāngāni
vāk prāṇash chakshuḥ shrotram
atho balam indriyāṇi cha sarvāṇi

"Everything is Brahman—this is the great teaching.
May I never deny Brahman.
May Brahman never deny me.
May there be no denial.
May there be no denial of me.
Let the great truths of the Upanishads
live in me, who delights in the Self."
Sarvaṃ brahmopanishadaṃ
māham brahma nirākuryāṃ
mā mā brahma nirākarot
anirākaraṇam astu
anirākaraṇaṃ me 'stu
tad ātmani nirate ya upanishatsu
dharmās te mayi santu
te mayi santu
Oṃ shāntiḥ shāntiḥ shāntiḥ

"This morning you will learn the secret of Brahman," said the āchārya. "And we will start with your questions." A murmur of excitement rippled through the group, because the children loved secrets. And even more than that, they loved to ask questions.

One of the older students eagerly started by asking, "*Kena*?" which means "by whom?"

"By whom does the mind think?" he asked his teacher. "By whom do the eyes see and the ears hear? By whose desire does life begin to move? By whose wish does a person begin to speak?"

The āchārya thought for a moment, looking at both the older and younger students. He decided to speak to the older students first.

"The answer is Brahman, which is wholeness," he said. "Brahman is the ear of the ear, and the mind of the mind. It is the breath of the breath, and the eye of the eye.

"Brahman cannot be expressed by speech, but it is that by which we are able to speak.

"Brahman cannot be thought by the mind, but it is that by which the mind thinks.

"Brahman cannot be seen by the eyes, but it is that by which the eyes see.

"Brahman cannot be heard by the ears, but it is that by which the ears hear."

The āchārya continued, "When someone knows Brahman, he finds truth, and that fortunate person becomes immortal. He lives forever. If, however, a person does not know Brahman, then his life is a great loss."

The teacher noticed that the younger students looked confused. They were saying with their eyes, "We don't understand what you mean!" So he decided to tell them a story. This is what he said—

At the beginning of time a vast ocean of milk covered the entire universe. The creamy ocean covered all that was. In the middle of this ocean towered a great mountain.

Vishnu, who maintains the creation, told the Devas how to churn the ocean so they could become immortal. "Take rare and precious herbs and throw them into the ocean," he said. "Use the mountain to stir the waters. Then you will receive the nectar of immortality, *amṛitam*."

The asuras also wanted the nectar of immortality, so they joined the Devas. Together they tied a serpent to the mountain, and pulled on the two ends. First the Devas pulled in one direction. Then the asuras pulled in the opposite direction. They turned the mountain harder and harder, faster and faster. Together they churned, like butter, the bottomless ocean of milk.

As they churned the ocean, many wonderful things emerged from the top of the mountain. Out came Dhanvantari, the physician of the gods, dressed in white robes and holding in his hand a golden vessel, called a *kalash*.

This vessel, everyone knew, contained within it the precious nectar of immortality, amṛitam. Both the Devas and the asuras badly wanted the amṛitaṁ, because they wanted to live forever. They wanted this more than they wanted anything else. Before anyone could stop them, the asuras suddenly snatched the kalash and raced across the sky with the prized amṛitam.

The Devas chased the asuras through the heavens. As they dashed across the heavens, four drops of nectar spilled out from the kalash. The four places where these drops fell later became the sites of large festivals, which have been celebrated for millennia in India. The largest of these festivals, held every twelve years in Allahabad, is called the *Kumbha Melā.*

The Devas fought a great battle with the asuras to win back the amṛitam. The battle lasted for thousands of years, but sadly, after some time, it appeared that the Devas were losing.

The sun and moon, the planets and stars watched anxiously from their homes high in the starry sky. They realized that the Devas could never win without help from the strongest power of all—Brahman. All of nature begged Brahman to rescue the Devas.

Just when it seemed that the Devas would surely lose, Brahman finally came to help. It is said that "Truth alone triumphs" (*Satyam eva jayate*), and so with truth on their side, the Devas captured the amṛitam. And that is how the Devas won the war with the asuras and gained immortality.

Unfortunately, instead of honoring Brahman for saving them, the Devas took credit for themselves. "We are so glorious!" they shouted. "We won the battle!"

Brahman saw their pride. He saw that they had forgotten who had won the war for them. So he decided to teach them a lesson.

Brahman appeared in front of the Devas in the form of a spirit. The Devas were baffled. Who was this spirit? What did he want?

They asked Agni to find out who this mysterious being was.

"Who are you?" the spirit asked as Agni came near.

"I am Agni Jātavedas," Agni replied.

"What power do you have?" the spirit asked.

"I am the great lord of fire," said Agni. "I can scorch everything. I can burn the entire world to ashes with the slightest thought."

Then the spirit placed a single blade of grass in front of Agni and said, "Burn this."

Agni howled and rushed in a red blaze toward the blade of grass. With all his strength he flared and flamed, trying to ignite it. But he could not kindle even a spark. Defeated, he returned to the other Devas.

"I could not find out who this spirit is before us," he said humbly.

The Devas then turned to Vāyu. They asked him to find out who the spirit was.

"Who are you?" the spirit asked as Vāyu came near.

"I am Vāyu Mātarishvan," Vāyu replied.

"What is your power?" Brahman (the spirit) asked.

"I am the great lord of wind," Vāyu said. "With a mighty gust of air I can blow away everything in the sky and on earth."

"Then blow this away," said the spirit, placing a single blade of grass in front of Vāyu.

Vāyu spun like a tornado and whirled around the blade of grass. He huffed and puffed a raging blast of air. But he could not move the blade of grass even a fraction of an inch. Defeated, he returned to the other Devas.

"I could not find out who this spirit is before us," he said, dejected.

The Devas then turned to Indra, their leader. "O Indra, find out who this spirit is!"

Mighty Indra in his shining robes and glittering crown went to the same spot. But the spirit had disappeared. In its place stood Umā, the pure and beautiful daughter of the Himālayas, the snow-capped mountains.

"Do you know who that spirit was?" Indra asked her.

"Yes, I know," Umā replied. "It was Brahman. It was because of Brahman that the Devas were victorious in battle. Because of Brahman the eyes see and the ears hear."

Indra now realized that it was Brahman who had won victory for the Devas. He and the other Devas finally admitted that they should not have been so proud. They praised Brahman for their victory and for their happiness.

❄❄❄

Having finished the story, the teacher asked his pupils if they understood. "Now do you know by whom the mind thinks?" he asked the youngest. "Do you know by whom the Devas achieved their victory? Do you know 'Kena,' do you know 'by whom'? Do you know by whom the eyes see and the ears hear? Do you know who is the eye of the eye and the breath of the breath?"

"Now we know by whom—by Brahman, the wholeness of the Self," answered a student. "By means of Brahman the eyes see and the ears hear. By means of Brahman the Devas received their strength."

The teacher smiled and nodded happily. "Brahman is the goal of all desires—it is the dearest of all," he said. "When the mind is settled, Brahman dawns like a flash of lightning, like the twinkling of the eyes.

"Whoever knows Brahman becomes dear to everyone. Whoever knows Brahman becomes immortal. Now you know the secret. You have been taught the secret teaching of Brahman."

The students left their seats and, one by one, quietly touched the feet of the teacher. Then the great āchārya stood and walked back to his cottage, his sandals whispering in the grass. The morning's lesson was over.

ब्रह्मविद् ब्रह्मैव भवति

Brahmavid Brahmaiva bhavati

The knower of Brahman is Brahman itself.

Muṇḍaka Upanishad, 3.2.9

The knower of Brahman is Brahman itself

CHAPTER SEVEN

Yāgyavalkya the Great Teacher

From the Bṛihadāraṇyaka Upanishad

In India there was once a great teacher named Yāgyavalkya. He was famous throughout the land for his knowledge and experience of Brahman. He lived in a glorious kingdom called Videha, where the people dwelled in perfect health and great happiness.

In Videha the rains came on time and the crops were plentiful. The markets overflowed with mangoes, bananas, oranges, and papayas. This was because the people lived pure lives and the king conducted many yagyas to bring goodwill to the kingdom.

One year the wealthy and generous king of Videha, King Janaka, decided to hold a large yagya. He invited all the wisest Brahmins (scholarly teachers) in the kingdom, including the most famous, Yāgyavalkya.

King Janaka himself was a wise person, and was in the habit of honoring the Brahmins by giving them generous gifts. On the day of his great yagya, as he looked at the learned teachers gathered before him, King Janaka suddenly wanted to know who among them was the greatest. Which star was the brightest?

To find out, King Janaka decided to offer a prize. He set aside one thousand cows from the royal herd. From the horns of each cow he hung ten gold coins.

"Most revered Brahmins," he said to the assembled sages, "whoever among you is the wisest, that person may drive home these cows."

No one moved. Each looked at the others in suspense. For sure, the prize was tempting, but who would dare claim the cows? Who would prove to be a superior Vedic scholar in such a meeting of the wise?

The air hung heavy and hot. It was so still they could hear the cows lowing in the distance. Finally, Yāgyavalkya stood up. He said to his young pupil, "Sāmashravas, my dear, drive home those cows." And the young boy took the cows away.

Suddenly everyone was in an uproar. All the other Brahmins cried out, "How can he declare himself the wisest?" To quiet the crowd, King Janaka's chief pandit, Ashvala, approached Yāgyavalkya.

"Are you indeed the wisest among us?" he asked Yāgyavalkya respectfully.

"I bow to the wisest among us," replied Yāgyavalkya, "but I really want those cows."

And so began a great debate to decide who was the greatest teacher, with King Janaka looking on.

Ashvala, Ārtabhāga, Bhujyu, and many other sages stood up to question Yāgyavalkya. Some asked questions about how to perform yagyas. Some asked about the senses, the objects of the senses, and the directions in space. Others asked about the nature of Ātmā (the Self). The debate continued all day, with all the pandits and scholars intent on hearing every word. Each person thought they would be the one to vanquish Yāgyavalkya, yet each was silenced by his brilliant answers.

The sun was slipping low in the sky when the great Vedic scholar Gārgī stood up. "Yāgyavalkya," she said, "My questions are sharp and pointed, like two arrows. What is above heaven, beneath the earth, and yet between the two? What is woven across the past, present, and future?"

"O Gārgī, it is Brahman," Yāgyavalkya answered without hesitating. "Brahman is above heaven, beneath the earth, and yet between the two. Brahman is the same in the past, present, and future.

"Brahman is neither short nor long. It is without taste, without smell, and without hearing. Yet at its command, the sun and moon rise in the east and set in the west, O Gārgī. It commands the endless cycles of the days and nights, the seasons and years.

"Brahman is unseen, but is the seer. It is unheard, but is the hearer. It is unthought, but is the thinker. There is no other seer than this, no other hearer than this, no other knower than this. In it everything is woven, like the warp and woof of cloth. Nothing exists but that (*Neha nānāsti kinchana*)," said Yāgyavalkya.

And so Gārgī, like the others, bowed to Yāgyavalkya. He asked if there were any more questions. No one else dared to speak.

"Whether one offers gifts in yagyas, like King Janaka, or sits still in meditation, the goal of life is to realize Brahman," Yāgyavalkya said.

Then all the wise sages declared Yāgyavalkya to be the greatest teacher, for he showed himself to be the most enlightened in the whole kingdom. King Janaka garlanded him with rose and jasmine blossoms.

And that is how Yāgyavalkya won the thousand cows.

Long after the debate, King Janaka continued to live in his jewel-studded palace with marble floors and sandalwood walls. Known far and wide for his dazzling wealth, King Janaka seemed to enjoy all the happiness life could offer. He and his wife had a beautiful and enlightened daughter named Sītā. The king was just and kind, loved by all the people.

King Janaka was not satisfied, however, with all the riches that life had to offer. He wanted to learn more about Brahman. At that time, no one learned from books. Students learned under the guidance of a teacher, an āchārya. So one morning, King Janaka called the wise Yāgyavalkya to his court. The king wanted to question him about Brahman.

As Yāgyavalkya entered the court, King Janaka, dressed in shining silk, stood up and smiled. He greeted Yāgyavalkya with hands folded in respect and offered him a high seat of honor.

"*Namaste!* (Greetings!)" said Yāgyavalkya.

"Namaste!" the king said. "All is well with you?"

"Yes, all is well, Your Majesty," Yāgyavalkya answered.

The king asked the attendants to bring water for Yāgyavalkya, and to garland him with jasmine.

"O Yāgyavalkya," the king asked, "for what purpose have you come? Have you come to answer my questions or have you come for more cows?"

"For both, Your Majesty!" answered Yāgyavalkya, his eyes twinkling.

Then Yāgyavalkya said, "If someone were going on a long journey, they would need a fast chariot or a strong ship. Like that, life itself is a long journey and one needs a good mind. You have such a mind, Your Majesty. You are wise to ask subtle questions."

And so the king began to question him.

"My dear Yāgyavalkya," began King Janaka, "by what light does a person see?"

"A person sees by the light of the sun, Your Majesty," Yāgyavalkya answered.

"But if the sun has set, by what light does a person see?" King Janaka asked.

"If the sun has set, a person sees by the light of the moon, Your Majesty," answered the great sage.

"If the sun has set and the moon has set, by what light does a person see?" King Janaka asked.

"If the sun has set and the moon has set, a person sees by the light of fire," answered Yāgyavalkya.

"If the sun has set, the moon has set, and the fire has gone out, by what light does a person see?" King Janaka asked.

"If the sun has set, the moon has set, and the fire has gone out, a person sees by the light of the Self."

"If the sun has set, the moon has set, the fire has gone out, and the Self has gone out, by what light does a person see?" King Janaka persisted.

"You ask too many questions!" said Yāgyavalkya. "For the Self never goes out. It never sets. It is eternal. It is the light that shines within the heart, by which a person may always see."

With this King Janaka felt satisfied, having understood that the Self within is the eternal light of Brahman.

"The knower of Brahman is calm, self-controlled, and patient," Yāgyavalkya said. "He sees the Self in all things and all things in the Self. This is the world of Brahman. You have attained this world, Your Majesty. You have attained Brahman."

"I give you the kingdom of Videha," King Janaka said. "I give myself as your servant."

Yāgyavalkya, however, not desiring the kingdom, accepted more cows as a gift and went home.

❋❋❋

Many years after teaching King Janaka, Yāgyavalkya decided to leave his life as a householder, and become a forest dweller.

At that time, the ideal length of life was said to be a hundred years, or a hundred long autumns. It was divided into four parts: the first twenty-five years were called *brahmacharya,* or student life; then came *gārhasthya,* householder life; then *vānaprasthya,* forest dweller life; and finally *sannyāsa,* retired life.

The ideal of Vedic living was for students to gain enlightenment by the time they were adults, and enjoy an entire lifetime of bliss, fulfillment, and success. Yāgyavalkya had achieved this ideal state of life.

Now in his latter years, Yāgyavalkya called his dear wife, Maitreyī, and said, "My beloved, it is my time of life to become a forest dweller. What would you like me to give to you before I leave? Would you like my wealth?"

"Even if I possessed the wealth of the whole world," Maitreyī asked, "would I become immortal?"

"No, my dear wife, you would not."

"Please, my dear husband," Maitreyī requested, "give me something that will make me immortal."

"You are truly dear to me, Maitreyī," said Yāgyavalkya. "I will give you knowledge, and this will bring you immortality."

He spoke the following words:

"It is not for the sake of the husband
that the husband is dear,
but for the sake of the Self
that the husband is dear.
Na vā are patyuḥ kāmāya
patiḥ priyo bhavati
ātmanas tu kāmāya
patiḥ priyo bhavati

"It is not for the sake of the wife
that the wife is dear,
but for the sake of the Self
that the wife is dear.
Na vā are jāyāyai kāmāya
jāyā priyā bhavati
ātmanas tu kāmāya
jāyā priyā bhavati

"It is not for the sake of the children
that the children are dear,

but for the sake of the Self
that the children are dear.
Na vā are putrāṇāṃ kāmāya
putrāḥ priyā bhavanti
ātmanas tu kāmāya
putrāḥ priyā bhavanti

"It is not for the sake of everything
that everything is dear,
but for the sake of the Self
that everything is dear.
Na vā are sarvasya kāmāya
sarvaṃ priyaṃ bhavati
ātmanas tu kāmāya
sarvaṃ priyaṃ bhavati

"The Self should be seen, heard,
contemplated, and realized.
O Maitreyī, when the Self is known,
then everything is known."
Ātmā vā are drashtavyaḥ shrotavyo
mantavyo nididhyāsitavyaḥ
Maitreyi ātmano vā are darshanena shravaṇena
matyā vigyānenedaṃ sarvaṃ viditam

After giving Maitreyī the precious knowledge of the Self, Yāgyavalkya asked his wife a question. "My dear, have you noticed that you cannot grasp the sound that comes from a drum?"

"Yes," answered Maitreyī.

"However, if you grasp the drum, you can create the sound that comes out," he said. "In the same way, you cannot catch the tunes that come out of a flute. But when you grasp the flute, you can play it, and thereby create the sounds that come from it.

"Like that, when you know the Self, Ātmā, which is the source of creation, you can create the life that you desire." And then Yāgyavalkya went to dwell in the forest to further develop his spiritual life. Yāgyavalkya had been an āchārya, a great teacher, teaching the true understanding of the Veda, pure knowledge, which he knew from his own experience. Through his teaching he had fulfilled his life's purpose and helped many, many others to reach the goal of life—the experience of infinite bliss—Brahman.

आत्मा वा अरे द्रष्टव्यः श्रोतव्यो
मन्तव्यो निदिध्यासितव्यः

Ātmā vā are drashtavyaḥ shrotavyo
mantavyo nididhyāsitavyaḥ

That Ātmā alone, that simplest form of awareness alone,
is worthy of seeing, hearing, contemplating, and realizing.

Bṛihadāraṇyaka Upanishad, 2.4.5

Out of bliss these beings are born, in bliss they are sustained, and to bliss they go and merge again

CHAPTER EIGHT

Bhṛigu Discovers the Nature of Brahman

From the Taittirīya Upanishad

One day long ago, a small boy named Bhṛigu sat in the shade of a palm tree with his father, Varuṇa. White clouds drifted slowly above fields of rice. Harvesters bent low, sickles flashing in the sun. Water buffalo waded in the shallow waters of the field.

Father and son sat quietly together. "Dear Father," Bhṛigu said, "teach me about Brahman."

"My dear son," Varuṇa said, "I am happy you want to know about Brahman. Do you see the clouds rolling across the sky? Do you see the growers harvesting the plump rice kernels? Do you see the water buffalo wading in the water?"

"Yes, Father," Bhṛigu said.

"All these are composed of matter, breath, sight, hearing, mind, and speech (*annaṃ prāṇaṃ chakshuh shrotraṃ mano vācham*)," his father said.

"That out of which these are born,
that in which they are sustained,
and that to which they go and merge again—

Yato vā imāni bhūtāni jāyante
yena jātāni jīvanti
yat prayanty abhisaṃvishanti

"Seek to know that, O Bhṛigu. That is Brahman. That is wholeness."

"Dear Father, how can I experience Brahman?" asked Bhṛigu.

"You will experience Brahman through meditation (*tapas*)," his father said. "Soon you will be old enough to receive your initiation, *Upanayana*. Then you will learn how to meditate."

Bhṛigu's heart swelled with joy at his father's words. One morning a few weeks later, Bhṛigu's family and friends gathered together for his initiation. Bhṛigu's mother cut his hair and set out new white clothes for him to wear after his bath.

When Bhṛigu entered the courtyard, the pandits were already chanting. As his mother watched with loving eyes, Bhṛigu approached his father, his hands folded to show respect. Together Bhṛigu and his father found a place of honor beside the orange-robed pandits, facing the sun. The ground was covered with marigold petals. Incense filled the air.

At the auspicious moment, Varuṇa gently placed the special cotton thread, woven in three strands, over Bhṛigu's left shoulder. He poured water from his own cupped hands into Bhṛigu's hands. He asked Bhṛigu to stand firm on a large rock, steadfast like a steady mind. Varuṇa said,

"All good you should hear from the ears.
All good you should see through the eyes."
Bhadraṃ karṇebhiḥ shṛiṇuyāma devāḥ
bhadraṃ pashyemākshabhir yajatrāḥ

Then Bhṛigu's mother and aunts fed him round, golden sweets, called *laddus*. Finally, after the initiation ceremony was over, Bhṛigu's father took him to a quiet place and taught him to meditate.

Bhṛigu closed his eyes. At once he felt a deep stirring within his consciousness. Now he knew he would discover the nature of Brahman. His father asked him to practice meditation every morning and afternoon.

That afternoon Bhṛigu waded across the rice field and sat next to a fast-moving stream. Soon he sank into silent meditation like an ancient Ṛishi (seer) sitting under an umbrella near the Ganges River.

After his meditation, Bhṛigu walked through the cool forest. He noticed the golden oriels spreading their bright yellow wings over the green trees. He felt the soft, warm earth under his bare feet.

"Maybe matter (*anna*) is Brahman," Bhṛigu thought. "All these beautiful things are composed of matter."

He thought,

"Out of matter these beings are born,
in matter they are sustained,
and to matter they go and merge again."

Annād dhy eva khalv imāni bhūtāni jāyante
annena jātāni jīvanti
annaṃ prayanty abhisaṃvishanti

The next day, after sweeping the kitchen, Bhṛigu asked his father, "Dear Sir, is it possible that matter is Brahman?"

"Yes, my dear son," Varuṇa answered, "matter is Brahman. And yet Brahman is more than that. Seek to know more about Brahman through meditation."

For several days Bhṛigu returned to the same spot, closed his eyes, and felt his mind become completely silent. As tamarind fruits dropped to the ground and monkeys scampered through banana trees, Bhṛigu sat peacefully in meditation by the stream.

After meditation, Bhṛigu noticed the peaceful morning breezes drifting over the water. As he breathed in, he smelled the cork tree flower, sweet like nutmeg. He thought about breath (*prāṇa*).

"Breath must be Brahman," he reflected. "After all, no one exists without breathing. Everyone depends upon breath."

He thought,

"Out of breath these beings are born,
by breath they are sustained,
and to breath they go and merge again."
Prāṇād dhy eva khalv imāni bhūtāni jāyante
prāṇena jātāni jīvanti
prāṇaṃ prayanty abhisaṃvishanti

Bhṛigu ran back to his father, full of excitement, but out of breath! "Dear Sir," he said, "is it possible that breath is Brahman?"

"Yes, you are correct, dear son," said Varuṇa quietly. "Breath is Brahman. And yet Brahman is more than that. Seek to know more about Brahman through meditation."

Bhṛigu knew that his father wanted him to find out the truth for himself. He knew that the knowledge he was seeking must come from within.

For several weeks Bhṛigu continued to meditate quietly every morning and evening. The silver-gray cranes dipped their long legs into the water and the snow-white egrets swooped for fish as Bhṛigu sat peacefully with closed eyes. Bhṛigu noticed that his mind was becoming more and more expanded. His mind felt playful like the forest animals, yet silent like the full moon.

One day after meditation he wondered, "Maybe mind (*manas*) is Brahman. When our minds are drowsy, we rest in the night, feeling peaceful and content. When our minds are refreshed, we awaken in the morning, feeling loving and creative."

He thought,

"Out of mind these beings are born,
in mind they are sustained,
and to mind they go and merge again."
Manaso hy eva khalv imāni bhūtāni jāyante
manasā jātāni jīvanti
manaḥ prayanty abhisaṃvishanti

That afternoon, after bringing home fruit, sugar, and nuts from the market, Bhṛigu said to his father, "Sir, I think that mind might be Brahman."

"Yes, dear son, mind is Brahman," Varuṇa answered. "And yet Brahman is more than that. Seek to know more about Brahman through meditation."

As time passed, Bhṛigu's responsibilities at home and in the village increased. He became more skillful as he milked the cows, fixed the roof, and helped his brothers and sisters with their chores.

As he continued to meditate and his mind became more and more settled inside, Bhṛigu became aware of the orderly patterns of nature—the cycles of the seasons and the movement of the stars. Even the orange butterfly seemed to be a picture of perfect order and intelligence.

Then Bhṛigu thought, "Intelligence *(vigyāna)* is Brahman."

He thought,

"Out of intelligence these beings are born,
through intelligence they are sustained,
and to intelligence they go and merge again."
Vigyānād dhy eva khalv imāni bhūtāni jāyante
vigyānena jātāni jīvanti
vigyānaṃ prayanty abhisaṃvishanti

Soon Bhṛigu sat next to his father. "Dear Father," he said, "could it be that intelligence is Brahman?"

Again his father answered, "My dear son, you are right. Intelligence is Brahman, and yet Brahman is more than that. Seek to know more about Brahman through meditation."

For many weeks Bhṛigu continued to meditate morning and evening. Gradually he noticed that he was enjoying his meditations more and more. A peaceful feeling blossomed in his heart. He felt clear and alert after meditation. Joy and happiness filled every moment of the day. The colored feathers of the birds, the faint rays of the sun peeking through the forest, the gurgling of the stream—all filled him with enchantment.

Bliss seemed to live in him and yet was all around him. It filled every cell of his body and every atom of the cashew nut tree, the myna bird, the water buffalo. Bliss was all that he was and bliss was all that he could see.

And then suddenly he knew that bliss (*ānanda*) is Brahman. Bliss is perfect harmony, eternal joy, perfection, and contentment.

Bhṛigu thought,

"Out of bliss these beings are born,
in bliss they are sustained,
and to bliss they go and merge again."
Ānandād dhy eva khalv imāni bhūtāni jāyante
ānandena jātāni jīvanti
ānandaṃ prayanty abhisaṃvishanti

"The whole world moves in bliss," he thought. "Even the sun and moon stay in their orbits because of bliss."

He burst into a beautiful song: "O wonderful, O wonderful, O wonderful! I have fathomed the entire universe. I am bliss. I am Brahman (*Ahaṃ Brahmāsmi*)."

As he greeted his father that evening, he didn't have to say anything. Varuṇa could see from his son's face that at last Bhṛigu knew the true nature of Brahman. Bhṛigu was radiating like the sun.

"My dear son, you have found Brahman, I see," Varuṇa said as he smiled and embraced him. "You have discovered that Brahman is bliss!"

"Yes, Father, I have found myself. I am bliss. I am Brahman. I am wholeness."

"Whoever knows the wisdom that you know, O dear Bhṛigu," said his father, "he is established in the Absolute, beyond space (*parame vyoman*).

"Whoever knows what you know in your heart, stands firm.

That person enjoys the entire creation.
That person enjoys wealth.
That person enjoys their family.
That person enjoys the love of all.
That person enjoys wisdom.
That person is a knower of Brahman."

And that is how Bhṛigu discovered the wholeness of Brahman, the totality, and grew up to become a famous Ṛishi of the Veda.

आनन्दाद्ध्येव खल्विमानि भूतानि जायन्ते
आनन्देन जातानि जीवन्ति
आनन्दं प्रयन्त्यभिसंविशन्ति

Ānandād dhy eva khalv imāni bhūtāni jāyante
ānandena jātāni jīvanti
ānandaṃ prayanty abhisaṃvishanti

Out of bliss these beings are born,
In bliss they are sustained,
And to bliss they go and merge again.

Taittirīya Upanishad, 3.6.1

CHAPTER NINE

Bālāki the Proud Teacher

From the Bṛihadāraṇyaka Upanishad

The most sacred river in India is the Gangā, the River Ganges—the Granter of Wishes. Along the Ganges many holy cities can be found, but the most holy city is Vārāṇasī. In ancient times it went by the name of Kāshī. Kāshī was the most beautiful place in the world. Its lakes and gardens surpassed even the heavens in beauty. The whole city was said to float in the sky, between heaven and earth.

Early one morning, in the misty darkness before dawn, a Brahmin named Bālāki walked quietly through the narrow lanes of Kāshī in his orange robes. He was called Proud Bālāki because he took great pride in his ability to give speeches in a grand and lofty manner. He was well traveled, and was known far and wide for his silver-tongued oratory.

Just as Bālāki was a famous Sanskrit scholar, Kāshī was a famous center of Sanskrit learning. The entire population devoted themselves to knowledge. Sanskrit pandits recited the Vedas and performed yagyas. Jyotish pandits made predictions for the future. Actors staged magical plays of the great epics, the Rāmāyaṇa and Mahābhārata. Kāshī was called the City of Light because it was lit from within. It was a city devoted to the light of knowledge.

As Bālāki walked through the city, he passed many shrines and temples. The most famous was the Vishvanātha Temple to Lord Shiva, its pointed roof made of solid gold. Countless pilgrims traveled on foot to Kāshī to worship there and bathe at dawn in the Ganges, the River of Life.

By the time Bālāki reached the Ganges, it was no longer dark and he was no longer alone. As the pale sun rose, Bālāki was joined by a flood of pilgrims coming to the river for their morning bath. Everyone walked down long flights of stone steps, called *ghāts,* to the river. There Bālāki boarded a small boat and crossed the river, as birds flew through the rising mist.

On the opposite shore, Bālāki reached the glorious palace of King Ajātashatru, the Rājā of Kāshī. This king, he knew, had no enemies. Bālāki entered the court. Surrounded by ministers and attendants, the humble King Ajātashatru greeted Bālāki with hands folded in respect, for that was the customary way for a king to honor a teacher. Bālāki walked proudly up to the king.

"Your Royal Majesty," he announced in a booming voice for all to hear, "I would like to teach you about Brahman."

"Thank you, kind Sir," the soft-spoken King Ajātashatru replied. "I would be delighted to receive your teaching. I will give you a thousand cows. Let us begin the instruction."

"Yes, Your Majesty," Bālāki said. "I will tell you about Brahman, the totality. We will begin this very moment. I understand Brahman as the sun, *Āditya*. This is the highest reality."

"Kind Sir," the intelligent king responded in a surprised tone, "I know that the sun gives radiance to all beings, creating a sparkle in their eyes. If I became the sun, then I too would give radiance to all beings. Certainly this is very good. But it is not the highest reality, which is Brahman."

Undaunted, Bālāki said, "You are right, Your Lordship. I understand that the moon, *Soma*, is Brahman."

At this King Ajātashatru was even more surprised. But still he replied modestly, "I know that the moon is the white-robed Soma, which stimulates the mind and is the essence of food. Certainly this is very good. But it is not the highest reality, which is Brahman."

Bālāki thought for a moment, and then said with authority, "I understand that lightning is Brahman."

To this the wise Ajātashatru replied, "I know that lightning is brilliance, which is found in the brightness of the skin. Certainly this is very good. But it is not the highest reality, which is Brahman."

Proud Bālāki again thought for a moment and said, "I understand that space, *ākāsha,* is Brahman."

To this King Ajātashatru replied patiently, "I know that space is full and without motion. It is found in the center of the heart, the source of love. Certainly this is very good. But it is not the highest reality, which is Brahman."

"I understand that air, *Vāyu,* is Brahman," said Bālāki, not to be silenced.

To this King Ajātashatru replied, "I know that air is found in the vital breath, and that it conquers all—when one's breath is settled, one has no enemies. Certainly this is very good. But it is not the highest reality, Brahman."

"I understand that fire, *Agni,* is Brahman," said Bālāki.

King Ajātashatru replied, "I know that fire is tolerant. It is found in a warm and sweet voice. Certainly this is very good. But it is not the highest reality, which is Brahman."

"I understand that water, *āpas,* is Brahman," said Bālāki.

"I know that water is harmonious and creates an agreeable nature," King Ajātashatru replied. "Certainly this is very good. But it is not the highest reality, Brahman."

At this point, proud Bālāki fell silent.

"Is that all?" King Ajātashatru asked respectfully.

"That is all," said Bālāki. For the first time in his life, he had nothing more to say.

"Kind Sir," the king said gently, "everything that you have told me is true. Brahman is all these things. But each one is only a part of Brahman. None alone is the totality. I do not think this teaching is enough to know Brahman."

Bālāki suddenly saw the king's wisdom. "Your Majesty," he said humbly, "will you take me as your student? I think I can learn what Brahman is from you. Please grant me this wish."

This odd request surprised everyone in the court. "Bālāki came here as a teacher and now he wants to be a student!" they murmured to each other.

At that time in India, the various functions in life were clearly defined. Four roles were said to come from the universal Being, called *Purusha,* the administrator of the entire universe. From his head came the *Brahmins,* the teachers and scholars. From his arms came the *Kshatriyas,* the warriors and kings. From his thighs came the *Vaishyas,* the merchants and farmers. From his feet came the *Shūdras,* the laborers and sweepers.

"It is very unusual for a Brahmin like Bālāki to ask a Kshatriya like me to teach him," said King Ajātashatru. However, the compassionate king saw that Bālāki had lost his pride. And so the king said, "Yes, I will teach you what Brahman is. Come with me."

Then the king rose, took Bālāki by the hand, and led him into the royal chambers. There they came upon a member of the royal household who was sleeping.

"Let us see if we can wake him up," the king whispered to Bālāki. The king raised his voice, "O Great One, get up!"

The man slept peacefully on.

"O Radiant One, cease your slumbers!" the king cried still louder.

Still no movement.

"O Soma, awaken!" Bālāki and the king shouted.

Nothing.

Finally, the king gently rubbed the man's hand, and he woke up.

"Where did this person go as he slept?" the king questioned Bālāki. "From what place did he return?"

"I do not know, Your Majesty," said Bālāki.

"When asleep, this man had withdrawn his senses," said the king. "His speech was withdrawn, his seeing was withdrawn, his hearing was withdrawn, and even his mind was withdrawn. While his intelligence sleeps, he resides in the space within the heart."

"What happens when he is dreaming?" Bālki asked.

"It is as if he becomes what he dreams," said King Ajātashatru. "Whether a great king or a worthy teacher, he dreams according to his past actions. The intelligence which sleeps, the intelligence which dreams—that is Brahman."

Then the king said, "Come Bālāki. Let us enjoy the pure river Ganges." Together they boarded the large royal boat and sat in soft red chairs. As they floated by Kāshī's famous ghāts, they saw people taking their baths, cleaning their clothes and making offerings to the river with cupped hands.

"Would you like to bathe at the royal ghāt?" the gracious king asked Bālāki.

"I would be honored," said Bālāki.

The boatmen rowed to the royal bathing ghāt. Bālāki and the king walked up the steps.

"Look, Bālāki," the king said. "Do you see that spider?"

"Yes," said Bālāki, "I see the spider moving along its web."

"We are like the spider," said the king. "We weave our life, and then move along in it. We are like the dreamer who dreams and then lives in the dream.

"This is true for the entire universe. That is why it is said, 'Having created the creation, the Creator entered into it' (*Tat shṛishtvā tad evānuprāvishat*).

"This is true for us. We create our world, and then enter into that world. We live in the world that we have created.

When our hearts are pure, then we create the beautiful, enlightened life we have wished for."

As they talked, Bālāki and the king watched the pandits build a fire for the mid-day yagya. "Do you see the fire?" the king asked.

"Yes, I see the fire."

"Do you see the sparks scattering from its center?" asked the king.

"Yes, Your Majesty, I see the sparks," said Bālāki.

"The world is like the sparks," said the king. "The world is a part of Brahman just as sparks are a part of the fire. The sun, moon, and lightning are a part of Brahman, like the sparks in a fire. Our sleep, our dreams, and our whole life are a part of Brahman, like the sparks in a fire.

"Brahman is the totality. It is wholeness. And Brahman is known by knowing the Self. This is the truth of truths."

Then Bālāki realized how simple the understanding of Brahman was. He was Brahman! Brahman was himself. By knowing himself, he could know everything and do anything. Bālāki humbly thanked the good king for giving him the supreme knowledge of Brahman—there by the Ganges, the River of Life, the Granter of Wishes.

अहं ब्रह्मास्मि

Ahaṃ Brahmāsmi

I am Totality.

Bṛihadāraṇyaka Upanishad, 1.4.10

CHAPTER TEN

Shvetāshvatara Teaches about Brahman

From the Shvetāshvatara Upanishad

High in the Himālayas, the Abode of Snow, small crystals of ice sparkle in the clear and fresh air. As the ice melts, it trickles down the mountain, and becomes a small stream. The stream becomes a creek, and the creek swells into a surging river. This is the most sacred river in India, the Ganges—the Stream of Nectar.

The Ganges flows from heaven to earth through deep gorges between the towering mountains. Here the air is silent, cold, and thin. The blue sky peeks from behind white and silver mountain tops. Pine trees hide many small caves. For many centuries, sitting in these caves high in the mountains, Ṛishis have practiced their meditations in silence.

One such Ṛishi in ancient times was Shvetāshvatara. For years the sage sat among the waterfalls, far above the torrent of the Ganges. From his small cave he heard the rapping of the red-headed woodpecker, trying to disturb his meditations. He heard the hum of the cicada echoing up the canyon past ferns and fields of wildflowers.

Shvetāshvatara felt satisfaction when it rained and contentment when it snowed. With joy he saw the distant peaks turn pink with the first rays of the sun. Filled with happiness, at peace with himself, he was half on earth and half in heaven.

Through his meditations, Shvetāshvatara had come to know the Self. He had become one with all he saw—mountains, misty clouds, rainbows, birds, and flowers. He was a fully enlightened sage. And yet he thought, "I am completely happy, but there is something I must do."

Some years later, Shvetāshvatara's cave was empty, but the Ganges, the River of Life, continued to flow. On its banks, the pointed roofs of shrines and āshrams touched the sky. One such Vedic school rested high on the cliffs above the river, at the foot of the snow-capped Himālayas. Monkeys scampered along the lime-green branches of the *pippala* tree in the quiet courtyard. The vanilla scent of frangipani blossoms mingled with the fragrance of sandalwood incense.

One day, as on every other day, Shvetāshvatara, the wise sage of the Himālayas, sat under the tree in the early afternoon with a group of students. Shvetāshvatara had left his cave to become a teacher. He was teaching his students Brahma Vidyā, the knowledge of Brahman, totality of life.

Shvetāshvatara began with an invocation,

"Let us be together.
Let us eat together.
Let us be vital together.
Let us be radiating truth,
radiating the light of life.
Never shall we denounce anyone,
never entertain negativity."
Saha nāv avatu
saha nau bhunaktu
saha vīryaṃ karavāvahai
tejasvi nāv adhītam astu
mā vidvishāvahai

"Look up!" Shvetāshvatara said. "Do you see the two large birds sitting in the tree?"

"Yes, Sir, we see them," the students said.

"Do you see how one of them eats the sweet pippala berry, while the other looks on without eating?" Shvetāshvatara asked.

"We see this," the students answered.

"You are like these two birds—one part dynamic, the other part silent. These are the two sides of your life. One part of you is active, and another part is a silent witness," Shvetāshvatara said.

"Sir, which bird is Brahman?" one student asked.

"Brahman is both together," answered Shvetāshvatara. "That is why Brahman is called the totality. No one reaches the highest tip of the tree without knowing Brahman.

"If you know only the silent bird, then you enter into darkness. If you know only the active bird, then you enter into a still greater darkness. But when you know both together, then you overcome death and obtain immortality (*Avidyayā mṛityuṃ tīrtvā vidyayā amṛitam ashnute*)."

Shvetāshvatara asked, "Do you see the swan floating on the Ganges?"

"Yes, Sir, we see," answered the students.

"In this vast wheel of Brahman, which creates all things, and in which all things rest," Shvetāshvatara continued, "the swan flutters about, thinking, 'My silent Self and the world are two different things.'

"But, in truth, they are not different. Some wise people see them as one, and those people rest in evenness. They gain eternal happiness."

"Please, Sir, tell us more about Brahman," another student asked.

Shvetāshvatara replied,

"He is one.
He is without form.
Through his own great power,
and for his own unfathomable purpose,
he creates countless forms.
He creates the world,
and, at the end of time,
gathers it back into himself.
May he give us clear understanding.

Ya eko 'varṇo
bahudhā shakti-yogād
varṇān anekān nihitārtho dadhāti
vi chaiti chānte vishvam ādau sa devaḥ
sa no buddhyā shubhayā saṃyunaktu

"He is Agni, the fire.
He is Āditya, the sun.
He is Vāyu, the wind.
He is Chandramā, the moon.
He is Shukra, the pure.
He is Hiraṇyagarbha, the golden womb of creation.
He is Āpas, the water.
He is Prajāpati, the protector of life.
Tad evāgnis
tad ādityas
tad vāyus
tad u chandramāḥ
tad eva shukraṃ
tad brahma
tad āpas
tat prajāpatiḥ

"You are woman.
You are man.
You are son and daughter, too.
You are the old man stumbling along with a staff.
Once born, you are the face in every direction.

Tvaṃ strī
tvaṃ pumān asi
tvaṃ kumāra uta vā kumārī
tvaṃ jīrṇo daṇḍena vanchasi
tvaṃ jāto bhavasi vishvato-mukhaḥ

"You are the dark blue butterfly.
You are the green parrot with red eyes.
You are the cloud, filled with lightning.
You are the seasons and the seas.
You are without beginning.
You are everywhere.
And all that is born, is born from you."
Nīlaḥ patango
harito lohitākshas
tadid-garbha
ṛitavaḥ samudrāḥ
anādimat tvaṃ
vibhutvena vartase
yato jātāni bhuvanāni vishvā

"Please tell us how to reach Brahman," another student asked.

"Through the practice of Yoga," Shvetāshvatara answered. "Through the settled mind."

"Please, Sir, how should we practice Yoga?" the student asked.

"Sit in a clean, level place that is beautiful," Shvetāshvatara said. "Let it be near the sound of water, and free from pebbles and fire. Let it be quiet, and protected from the wind.

"As you dive within, your mind will become settled, like the driver of a chariot yoked to good horses. As your mind becomes more calm, your breath will become lighter, and your body will become still and steadfast, like a rock.

"There you will find the Self, the state of union. There you will know the auspicious Brahman, the Veda, and attain unending peace. There you will follow the path of the sun."

"Please, Sir, when we start the practice of Yoga, what can we expect?" another student asked.

"The beginning fruits of Yoga are lightness, good health, steadiness, clearness of complexion, and a pleasing voice," Shvetāshvatara said.

"Just as a mirror shines bright once it has been cleaned of dust," he continued, "so those who have seen the Self shine in mind and body. They are always and forever filled with happiness.

"It would be easier to roll up the entire sky into a small cloth than it would be to obtain true happiness without knowing the Self. Only by knowing the Self does one become immortal. There is no other path."

And this is how the wise sage, Shvetāshvatara, after realizing the Self through meditation and the grace of God, spoke about Brahman as the supreme, the pure, the undivided—far above the Ganges, and far below the Himālayas, in India, the Land of the Veda, the land of knowledge.

वेदाहमेतं पुरुषं महान्तम्
आदित्यवर्णं तमसः परस्तात्
तमेव विदित्वातिमृत्युमेति
नान्यः पन्था विद्यतेऽयनाय

Vedāham etaṃ purushaṃ mahāntam
āditya-varṇaṃ tamasaḥ parastāt
tam eva viditvātimṛityum eti
nānyaḥ panthā vidyate 'yaṇāya

I know the Veda, the great totality,
radiant as the sun, beyond darkness.
Those who know that become immortal.
There is no other path.

Shvetāshvatara Upanishad, 3.8

I know the Veda, the great totality,
radiant as the sun, beyond darkness
Those who know that become immortal
There is no other path

CHAPTER ELEVEN

How Creation Began

From the Aitareya Āraṇyaka
and the Aitareya Upanishad

Once in ancient India a boy and his father were sitting on the edge of a peaceful lake as the morning rays of the autumn sun met the snowcapped mountains, high above the white clouds. Willow trees and meadows of yellow flowers reflected in the ripples of the blue lake. White lotus flowers that opened only at night floated peacefully on the water.

The boy's name was Mahīdāsa Aitareya. Mahīdāsa was learning the Veda from his father, as he did every morning.

Today, Mahīdāsa was learning how to pronounce the Sanskrit alphabet. His father said, "*a ā, i ī, u ū,*" and Mahīdāsa repeated "*a ā, i ī, u ū.*" Over and over they practiced, Mahīdāsa repeating after his father.

Finally his father said, "Mahīdāsa, my dear, do you have any questions?"

"Yes, father," said Mahīdāsa. "Where does Sanskrit come from? Who created the Sanskrit alphabet?"

His father thought for a minute, and then said, "Sanskrit is called the language of nature, the language of Veda, *Vedavāṇī,* because no one created it. In its name (*nāma*) is the form (*rūpa*), and so it is perfected (*saṃskṛita*). It is the speech of nature. There is a story, however, about how the Sanskrit alphabet came to be known by humans."

And then Mahīdāsa's father told him the following story.

"Lord Shiva, called Natarāja, the Lord of Dance, performed a great dance in the center of the universe, which is located in the center of the heart. Shiva was the dancer, the dance, and the audience. With bells ringing and flames encircling him, he danced, his braided hair whirling as he spun 'round and 'round. In his left hand he held the flame of destruction and in his right hand he held the drum of creation, called *damaru*. He swung the drum from side to side, measuring the rhythm of the dance.

"From the pulse of that small drum came forth the Sanskrit alphabet, and from those sounds came forth all forms. The alphabet was heard by the sage Pāṇini. He recorded the cognition in fourteen short verses, called the *Shiva Sūtras*. Pāṇini placed these sūtras in the beginning of his Sanskrit grammar."

"Dear father," said Mahīdāsa, "of these letters of the alphabet, which is the most important?"

"The first is most important," Mahīdāsa's father said. "And that sound is 'A.' In 'A' is contained the entire alphabet, and from 'A' comes many sounds and also many forms. It is said,

> " 'A' is the whole of speech.
> Manifesting as sibilants and consonants,
> it becomes many different forms."

Akāro vai sarvā vāk
saiṣā sparśoṣmabhir vyajyamānā
bahvī nānā rūpā bhavati

"How is the alphabet divided?" asked Mahīdāsa.

"The alphabet, called *akshara samāmnāya,* or 'recitation of letters,' is divided into three parts," his father said.

"First there are the vowels, then the consonants, and then the sibilants ('s' sounds), which lie between the vowels and consonants. Each of these three is expressed in a different way:

"The vowels are expressed as Ātmā, the Self.
The sibilants are expressed as the breath.
And the consonants are expressed as the body.
Yo ghoshaḥ sa ātmā
ya ūshmāṇaḥ sa prāṇaḥ
yāni vyanjanāni tachchharīram

"It is said that at the beginning of creation the vowels, sibilants, and consonants formed the three realms:

"The vowels formed the celestial.
The sibilants formed the atmosphere.
And the consonants formed the earth."
Divaḥ svarā
antarikshasyoshmāṇaḥ
pṛithivyā rūpaṃ sparshāḥ

"Dear father," asked Mahīdāsa, "how should I pronounce each of these?"

His father said,

"Pronounce all vowels resonant and strong.
Pronounce all sibilants open, without slurring or sliding.
Pronounce all consonants slowly, without blending them.
Sarve svarā ghoshavanto balavanto vaktavyāḥ
sarva ūshmāṇo 'grastā anirastā vivṛitā vaktavyāḥ
sarve sparshā leshenānabhinihitā vaktavyāḥ

"If someone should criticize your vowels,
tell him, 'I have taken refuge in Indra.
He will answer you!'
Taṃ yadi svareshūpālabheta
indraṃ sharaṇam prapanno 'bhūvaṃ
sa tvā prati vakshyatīty enaṃ brūyāt

"If someone should criticize your sibilants,
tell him, 'I have taken refuge in Prajāpati.
He will smite you!'
Atha yady enam ūshmasūpālabheta
prajāpatiṃ sharaṇaṃ prapanno 'bhūvaṃ
sa tvā prati pekshyatīty enaṃ brūyāt

"If someone should criticize your consonants,
tell him, 'I have taken refuge in Mṛityu.
He will scorch you!' "

Atha yady enaṃ sparsheshūpālabheta
mṛityuṃ sharaṇam prapanno 'bhūvaṃ
sa tvā prati dhakshyatīty enaṃ brūyāt

Mahīdāsa and his father laughed and laughed. His father told him that they would continue with their next lesson tomorrow.

❄❄❄

The following morning, as the clear rays of the sun warmed the high mountains, Mahīdāsa and his father again sat beside the clear lake. The morning lesson was about to begin. Mahīdāsa's father faced the morning sun and said,

"May speech be established in my mind.
May my mind be established in speech.
O radiant one, be radiant unto me.
Fill my mind with the Veda.
Do not forsake my knowledge.
What I have known, may I know day and night.
Vāṇ me manasi pratishthitā
mano me vāchi pratishthitam
āvīr āvīr ma edhi
vedasya ma āṇīsthaḥ
shrutaṃ me mā prahāsīḥ
anenādhītenāhorātrān saṃdadhāmi

"I will speak of what is right.
I will speak of what is true.
May that satisfy me.
May that satisfy the teacher.
May I be satisfied.
May the teacher be satisfied.
May the teacher be satisfied."
Ṛitaṃ vadishyāmi
satyaṃ vadishyāmi
tan mām avatu
tad vaktāram avatu
avatu mām
avatu vaktāram
avatu vaktāram

Then he said to his son, "Mahīdāsa, my dear, do you have any questions today?"

"Yes, Sir. Could you tell me how creation began?"

"That is a good question, Mahīdāsa. There are many stories to explain the answer. Each of these stories is true, because truth has many sides and can be thought of in many ways. I will tell you one such story that illustrates the truth.

Once, long ago, before creation, before anything existed, there was nothing, except for one being, Ātmā. No one else was there. No one else winked.

Ātmā thought to himself, "Now let me create the worlds!"

And so he created the three realms—

In the first realm, the celestial,
he placed water.
In the second realm, the atmosphere,
he placed rays of light.
And in the third realm, the earth,
he placed Mṛityu, death.

Then he thought, "Here, then, are the three worlds. Now let me create the guardians of the worlds, the Devas."

Then, out of himself, he created the Devas and placed them in the ocean. But a problem developed with the Devas—they became hungry and thirsty. They said to Ātmā, "Please, Sir, give us an abode in which we may satisfy ourselves."

And so Ātmā created a cow for the Devas. The gentle cow had long eyelashes and brown eyes. She was patient, like the earth, and she mooed affectionately, like Kāmadhenu, the wish-fulfilling cow. Her horns curved in a graceful arc, and her udder overflowed with creamy milk, one of the seven sacred foods.

The Devas looked at the cow and then huddled in a circle, talking among themselves. Finally they said to Ātmā, "This is not enough!"

And so Ātmā created a horse. The swift stallion had a gait like a prince, prancing left and right, *clip-clop, clip-*

clop. He possessed auspicious marks (*lakshaṇa*) and all the correct proportions. Installed in him was a precious secret, the secret of the correct pronunciation of the Sanskrit letter '*ṃ*,' called *anusvāra*.

The Devas looked at the horse and then huddled in a circle, talking among themselves. Finally they said to Ātmā, "This is not enough!"

And so, out of himself, Ātmā created a person, a human. The human walked slowly, with clear eyes looking straight ahead. His hands were steady and his mind was tranquil. His forehead was wide, and his arms were long.

The Devas looked at the human and proclaimed, "Well done! This will do!"

"Very well," said Ātmā, "now you may enter into your respective abodes."

And then the Devas took their places in the human body, each selecting a location to dwell:

Agni entered the mouth
and became speech,
and through speech he became fire.

Vāyu entered the nostrils
and became breath,
and through breath he became air.

Sūrya entered the eyes
and became sight,
and through sight became the sun.

The quarters of space entered the ears
and became hearing,
and through hearing they became sound.

Chandramā entered the mind
and became the heart,
and through the heart became the moon.

Then Ātmā thought, "I have created abodes for the Devas. But I see they are still hungry. Now I will create food for them, so they may be satisfied. "

Ātmā created food for the Devas. But then still another difficulty arose. The human was unable to catch the food.

The human tried to catch the food with speech.
He was not able to do so.
Even if he had, his speech would not have been satisfied.

He tried to catch the food with his breath.
He was not able to do so.
Even if he had, his breath would not have been satisfied.

He tried to catch the food with his sight.
He was not able to do so.
Even if he had, his sight would not have been satisfied.

He tried to catch the food with his hearing and his mind.
Still, he was not able to do so.
Even if he had, he would not have been satisfied.

Ātmā looked at this person he had created, and saw that the human was not satisfied. Each of his senses was trying to satisfy itself alone. Ātmā wondered what could be done. He thought,

"If speaking is through speech,
if breathing is through the breath,
if seeing is through the eyes,
if hearing is through the ears,
if thinking is through the mind,
then who am I?

"I also will reside in this person," Ātmā thought to himself.

"Where should I enter?" he wondered. "Should I enter through the sole of the foot or through the crown of the head?"

Ātmā decided upon the head. Like a bolt of lightning, he burst through the crown of the head, and it opened like a lotus with a thousand petals. And so it is said, "Having created the creation, the Creator entered into it" (*Tat sṛishtvā tad evānuprāvishat*). Having entered the person, Ātmā resided in the center of the heart in a space the size of the thumb (*Angushtha-mātraḥ purusho 'antarātmā sadā janānāṃ hṛidaye sannivishtaḥ*).

Then Ātmā caused himself to be seen by the human. The person saw the Self, Ātmā, and he woke up, as if for the first time. Then his senses served Ātmā, instead of trying to satisfy themselves.

The human came to see that the Ātmā in himself was the same Ātmā who had created heaven and earth. The Ātmā in himself was the same Ātmā who had created the sun and moon, the air and water. This was the same Ātmā who had created the cow and horse. This was the same Ātmā who had created man. Then the human was satisfied. He thought, "This intelligence is Brahman" (*Pragyānaṃ Brahma*).

The human understood that the entire universe was in the silence of his own Self, and that when he knew himself, he knew everything. With that he soared upward with joy. He fulfilled all his desires and became immortal in the world of heaven (*Svarge loke sarvān kāmān āptvā amṛitaḥ samabhavat*). Yes, he became immortal in the world of heaven.

Mahīdāsa's morning lesson had ended. He touched his father's feet in respect and thanked him for the lesson. As Mahīdāsa tended the goats on the hillside, he practiced the Sanskrit letters he had learned, and he thought about his father's words. He thought about how the universe is located in the center of his heart. He looked forward to tomorrow's lesson by the lake, and thought of new questions to ask his beloved father.

प्रज्ञानं ब्रह्म

Pragyānaṃ Brahma

Fully awake self-referral dynamism (of the universe)
born of the infinite organizing power of pure knowledge, the
Veda—fully awake totality of the individual consciousness
is Brahman, which comprehends the infinite dynamism
of the universe in the infinite silence of the Self.

Aitareya Upanishad, 3.1.3

CHAPTER TWELVE

King Ashvapati and the Universal Self

From the Chhāndogya Upanishad

In the ancient kingdom of Kuru, several Brahmins lived near to each other. Their names were Prāchīnashāla, Satyayagya, Indradyumna, Jana, and Budila. These men were all great householders (*mahāshāla*) and great scholars (*mahāshrotriya*).

At one time, their land was invaded by locusts, and the crops withered in the fields. At the request of the king, the Brahmins of the kingdom performed a large yagya to restore balance in nature. The yagya was successful, and the people once again harvested bountiful crops. All the Brahmins in the kingdom were honored with gifts of cows, gold, homes, and land.

Every afternoon the six wealthy scholars would assemble in the spacious gardens between their homes to debate the true nature of Ātmā, the Self.

"Ātmā is one, like the sea," one scholar would say.

"No, Ātmā is many, like the trees," another would answer.

"Ātmā is unmoving, like the pole star," one scholar would say.

"No, Ātmā moves, like the sun," another would answer.

"Ātmā is unbounded, like the sky," one scholar would say.

"No, Ātmā is bound by past action, like the earth," another would answer.

They talked together like this for some time, but were unable to come to an agreement about the true nature of Ātmā. Each considered his own description to be true and could not understand how a seemingly opposite description could also be true.

One day Prāchīnashāla said, "Learned scholars, I have heard that there is a great teacher in the land of Panchāla who has traveled widely and has spent much time studying the Veda. Perhaps he knows the true nature of Ātmā. His name is Uddālaka Āruṇi, and he lives with his son, Shvetaketu. Let us go to him and ask him to settle our debate."

And so they decided to leave their kingdom of Kuru and journey to Uddālaka's village in the land of Panchāla.

"Please, Sir, would you resolve our debate about the nature of Ātmā?" they asked when they reached the home where Uddālaka lived.

Uddālaka listened to their arguments carefully. He thought, "These are all great scholars, and yet they have not been able to resolve among themselves the true nature of

Ātmā. How can I settle their dispute? They will continue to debate the points, with no end."

"Learned scholars," he said, "I suggest that we go together to see King Ashvapati. He has fathomed the universal nature of Ātmā. He will be able to settle the discussion."

And so the six wise scholars left Panchāla and journeyed to the kingdom of Kekaya, where King Ashvapati lived. King Ashvapati's beautiful daughter, Kaikeyī, was married to the famous King Dasharatha of Ayodhyā, who was the father of Shrī Rāma.

King Ashvapati lived in an enormous white palace, with gilded domes and inlaid gold on heavy mahogany doors. Inside, fountains splashed on large crystals, surrounded by pools with ferns and waving palms. The marble floor felt cool to their feet as the learned scholars walked through the hall.

The king was seated on a high royal seat made of *udumbara* wood covered with reeds. He greeted the scholars with jasmine garlands and coconuts, and his attendants washed their feet. The learned guests were given woolen shawls embroidered with Sanskrit verses. They enjoyed a banquet that ended in sumptuous sweets.

"Learned scholars," the king said, "I can see that you are tired from your journey. Please rest for the night, and I will answer your questions when you feel fresh."

After traveling for so long, the six scholars were happy to rest in comfortable rooms with balconies. The next morning, they felt refreshed and went immediately to the royal court.

"Dear scholars," King Ashvapati said to them, "in my kingdom all are content with their lives and are generous towards their neighbors. All have clear minds and enjoy harmony with their families. The reason for our good fortune is that my people understand the true nature of Ātmā and our Brahmins perform many yagyas.

"Today the learned Brahmins in our kingdom will perform a great yagya to bring wealth to the kingdom. For this performance they will receive many spiritual and material benefits, and they will also receive many gifts as *dakshinā* from me. I will give each Brahmin a thousand *droṇas* of corn, ten *bhāras* of molasses, twenty-one jars of ghee, and a thousand *palas* of gold. As much as I give to them, I will give to you."

With that the king invited his guests to attend the yagya. The scholars were happy to witness the yagya and receive the kind gifts from the king. As they watched the white smoke drift upward from yagya fires, the scholars understood why King Ashvapati's kingdom was so prosperous.

Afterward they thanked him for the gifts and said, "Your Royal Majesty, we have traveled far to ask you questions about the universal nature of Ātmā. Now we are eager to learn from you, for we see from your kingdom that there are, indeed, many great benefits to knowing the nature of Ātmā."

"Yes, I will tell you," the king said. "But first, please rest one more night."

And so the scholars stayed yet another night. The next morning, feeling fresh and vital, they went immediately to the royal court.

King Ashvapati began by asking his pandits to chant from the Vedas. The pandits recited from the Sāma Veda in a slow melodious tune, like the melody chanted by the Devas themselves at the dawn of creation. The recitation continued for a long time, and all listened with their eyes closed. The six scholars were so lost in contentment that they almost forgot their questions!

But it was the king who would question them first. In the silence after the recitation, the king examined the faces of his learned guests.

"My dear Prāchīnashāla," the king began,"what do you consider to be Ātmā?"

"I consider the sky (*diva*) to be Ātmā, Your Majesty," said Prāchīnashāla.

"What you consider to be Ātmā is that aspect of the universal Self that creates splendor (*sutejas*)," said the king. "As a result of this understanding, you and your family enjoy *soma,* the ambrosia of immortality. This nourishes you, this is dear to you. This nourishes them, this is dear to them. Such a person thinks he has understood Ātmā.

"This, however, is only the head of Ātmā. O Prāchīnashāla, your head would have fallen off if you had not come to me!" said the playful king.

Then the king turned to Satyayagya and said, "O Satyayagya, what do you consider to be Ātmā?"

"I consider the sun (*Āditya*) to be Ātmā, Your Majesty," said Satyayagya.

"What you consider to be Ātmā is that aspect of the universal Self that illumines all forms (*vishva-rūpa*)," said the King. "As a result of this understanding, you and your family enjoy delights in all forms—chariots, servants, and gold. These nourish you, these are dear to you. These nourish them, these are dear to them. Such a person thinks he has understood Ātmā.

"This, however, is only the eye of Ātmā. O Satyayagya, you would have become blind if you had not come to me!" said the king with a smile.

Then the king turned to Indradyumna and said, "O Indradyumna, what do you consider to be Ātmā?"

"I consider air (*Vāyu*) to be Ātmā, Your Majesty," said Indradyumna.

"What you consider to be Ātmā is that aspect of the universal Self that shapes different paths (*pṛithag-vartman*)," said the king. "As a result, rows of chariots laden with gifts come to you and your family from different paths like the wind. These nourish you, these are dear to you. These nourish them, these are dear to them. Such a person thinks he has understood Ātmā.

"This, however, is only the breath of Ātmā. O Indradyumna, you would have lost your breath if you had not come to me!"

Then the king turned to Jana and said, "O Jana, what do you consider to be Ātmā?"

"I consider space (*ākāsha*) to be Ātmā, Your Majesty," said Jana.

"What you consider to be Ātmā is that aspect of the universal Self that produces abundance (*bahula*). As a result, you and your family enjoy abundant children and wealth.

This nourishes you, this is dear to you. This nourishes them, this is dear to them. Such a person thinks he has understood Ātmā.

"This, however, is only the body of Ātmā," said the king. "O Jana, your body would have crumbled if you had not come to me!"

Then the king turned to Budila and said, "O Budila, what do you consider to be Ātmā?"

"I consider water (*āpas*) to be Ātmā, Your Majesty," said Budila.

"What you consider to be Ātmā is that aspect of the universal Self that generates wealth (*rayi*). As a result, you and your family enjoy wealth and prosperity. This nourishes you, this is dear to you. This nourishes them, this is dear to them. Such a person thinks he has understood Ātmā.

"This, however, is only the bladder of Ātmā," said the king. "O Budila, your bladder would have burst if you had not come to me!"

Finally, the king turned to Uddālaka and said, "O Uddālaka, what do you consider to be Ātmā?"

"I consider earth (*pṛithivī*) to be Ātmā, Your Majesty," said Uddālaka.

"What you consider to be Ātmā is that aspect of the universal Self that forms the foundation (*pratishtha*). As a result, you and your family are blessed with children and cattle. This nourishes you, this is dear to you. This nourishes them, this is dear to them. Such a person thinks he has understood Ātmā.

"This, however, is only the feet of Ātmā," said the king. "O Uddālaka, your feet would have withered if you had not come to me!"

Then the king spoke to all the wise Brahmins together. "Learned scholars, you consider Ātmā to be many things. And whatever aspect of Ātmā you exalt in—that aspect nourishes you, and that aspect is dear to you.

"Ātmā, however, is universal (*vaishvānara*). It stretches from earth to heaven. It is one, yet many. It moves, yet is unmoving. It is far, yet near. Ātmā is smaller than the smallest particle, yet bigger than the entire universe.

"No measuring rod can measure it. It is known only to itself. When you know Ātmā, you know the totality. Then everything nourishes you, and everything is dear to you.

"Know Ātmā as universal. Then you will fathom the whole of life, and you will live in all worlds.

"Now, of this universal Self,
the head, indeed, is the bright light;
the eye is every form;
the breath is of different paths;
the body is abundance;
the bladder, indeed, is wealth;
the feet, indeed, are the earth."
Tasya ha vā etasyātmano vaishvānarasya
mūrdhaiva sutejāḥ
chakshur vishva-rūpaḥ
prāṇaḥ pṛithag vartmātmā
saṃdeho bahulaḥ

bastir eva rayiḥ
pṛithivy eva pādau

Then the king told the learned scholars a story about six blind men who held scholarly debates among themselves about the true nature of an elephant.

"An elephant is small, like a rope," one blind man would say.

"No, an elephant is big, like a wall," another would answer.

"An elephant is round, like a tree," one blind man would say.

"No, an elephant is flat, like a piece of cloth," another would answer.

"An elephant is hard like a conch shell," one blind man would say.

"No, an elephant is soft like rubber," another would answer.

They talked together like this for some time, but were unable to come to an agreement about the true nature of an elephant. Each considered his own description to be true and could not understand how a seemingly opposite description could also be true.

Then along came a man who could see. He explained that the elephant was all of these together. He said,

"Now, of this elephant,
the tail is small, like a rope;
the body is big, like a wall;
the legs are round, like trees;
the ears are flat, like pieces of cloth;
the tusks are hard like conch shells;
and the trunk is soft, like rubber."

"Each blind man knew only a part of the whole," said the king. "Like this, the Self is wholeness. The Self is universal. When you truly know the Self, then you know everything. Then everything nourishes you, and everything is dear to you."

And that is how the six Brahmins, all great householders and great scholars, learned about the nature of the universal Self from King Ashvapati in the kingdom of Kekaya. They returned to their homes in the land of Kuru and Panchāla, and they gave their families and friends the gift that they had received from King Ashvapati—the knowledge of the universal Self.

एकमेवाद्वितीयम्

Ekam evādvitīyam

One reality without a second.

Chhāndogya Upanishad, 6.2.1

One reality without a second

CHAPTER THIRTEEN

Āshvalāyana Visits Brahma Loka

From the Kaivalya Upanishad

Āshvalāyana had one desire —to learn the essence of all wisdom, Brahma Vidyā. And so Āshvalāyana decided to visit Lord Brahmā, the Creator.

Āshvalāyana traveled north to Mount Meru, an enormous mountain made of pure gold, 160,000 leagues high. Jewels and herbs covered its slopes, and a gigantic rose-apple tree, called a *jambu* tree, shaded its southern side. Mount Meru was a mountain of beauty. From here the River Ganges flowed to earth. From here the sun and moon made their daily rounds.

On the summit of Mount Meru was a flat square extending for ten thousand *yojanas*. This was *Brahma Loka*, the celestial dwelling place of Lord Brahmā. As Āshvalāyana walked down its golden paths, he saw that Brahma Loka was filled with light, surpassing even the sun in brilliance. This place was also called *Satya Loka,* the Realm of Truth, because all who lived here knew the eternal truth of existence within themselves. In this place, *siddhas,* perfected beings, enjoyed peace and harmony.

Āshvalāyana came to a lake, called Lake Ara. As he gazed at the other shore, he instantly crossed over the lake and found himself there. Āshvalāyana remembered his past actions, both good and bad, and he saw them fly away, for like day and night, both are left behind when one reaches Brahma Loka. Āshvalāyana noticed that everyone here looked young, as if the moments of time had fled from them.

Āshvalāyana came to the center of Brahma Loka, to the great city of Brahmā, called *Manovatī*. It was a city of immortals (*devanāgarī*), with shining crystal buildings in orderly rows, surrounded by fountains and pools. In the middle of the city was the place of Brahmā, called the *brahmasthān*. It was a garden of splendor, and in its center sat the lotus-born Brahmā in deep meditation.

Next to Brahmā, sitting on a white lotus, was Sarasvatī. She held the *vīṇā*. Resting in a mango tree nearby was a green partridge (*chakora*), said to subsist on moonbeams. Its eyes turned red if it saw food that was not pure.

Even from a distance, Āshvalāyana could see kindness and compassion in Brahmā's face. As Āshvalāyana came closer, he felt the fragrance and flavor of Brahmā enter into him. As he came still closer, he felt the radiance and glory of Brahmā enter into him.

Brahmā, who even with eyes closed could see all things in all directions, acknowledged the presence of Āshvalāyana with a gesture of kindness. Āshvalāyana felt that Brahmā knew everything about him, even before he spoke.

Āshvalāyana bowed to Brahmā and said, "Revered Sir, please teach me the knowledge of Brahman. This is the highest wisdom, sought after even by the wise. It is the innermost secret, which frees the knower from impurities, whereby he attains Purusha, greater than the great."

Brahmā, the grandsire, said to him,

"Seek to know Brahman through faith,
devotion, meditation, and Yoga.
Not through action, children, or wealth,
but only by going beyond does one reach immortality.

"Beyond heaven, shining
in the depth of the heart,
it is attained by those
who are realized.

"Those ascetics who have grasped the essence
of the Vedānta wisdom, who have purified their nature
through the yoga of renunciation, are completely liberated
in the worlds of Brahmā, beyond time, beyond mortality.

"In a secluded place, seated in a comfortable posture,
purified, with head, neck, and body upright,
beyond duties, beyond all the senses,
having given devotion to the teacher,

"meditate and know the white lotus in the middle of the heart,
free from desire, pure, beyond thought, beyond sorrow,
calm, inconceivable, unmanifest, of infinite form,
blissful, peaceful, the immortal source of Brahmā.

"That has no beginning, no middle, no end.
It is far-extending, yet one. That is bliss consciousness, *chid-ānanda*. Without form, wonderful, it is the highest Lord, *Parameshvara*. It is the ruler. It is tranquility.

"Thus having meditated, sages go to the source of all, become the witness of all, and go beyond darkness.

"That is Brahmā. That is Shiva. That is Indra.
That is the imperishable, the highest, the ruler, the Self.
It is Vishnu. It is the vital breath. It is time, *kāla*.
It is light, *Agni*. It is the moon, *Chandramā*.

"That is all that was and all
that will be. That is eternal.
Knowing that one goes beyond death.
There is no other path to liberation.

"Seeing the Self in all beings,
and all beings in the Self,
one goes to the supreme Brahman,
not by any other cause.

"Having made the Self the lower firestick,
and sound the upper firestick,
through daily kindling of the flame of knowledge,
the wise person burns the bonds of ignorance.

"A person, completely deluded by illusion,
obtains a body and performs actions of every kind.
While awake, he is satisfied by various enjoyments,
such as women, food, and drink.

"While dreaming, the person experiences joy and sorrow
in worlds fashioned by his own illusion.
While sleeping, everything vanishes in darkness,
and he appears to be happy.

"And again, linked to the actions of his former birth,
the person wakes up, and again, he sleeps. He plays
in these three states, and creates everything in his world.
In him who is the vessel of bliss and undivided wisdom,
these three states are merged.

"From him are born the vital breath,
the mind, and all the senses.
From him are born space, air, fire,
water, and earth, the foundation of all.

"That which is the supreme Brahman,
the Self of all, the great abode
of the universe, subtler than subtle,
eternal—that you are; you are that.
Tat tvam eva tvam eva tat

"Diversity shines forth
in waking, dreaming, and sleeping.
Knowing 'I am Brahman,'
one is freed from all bonds.

"In the three states are the enjoyer (*bhoktṛi*),
the enjoyed (*bhogya*), and enjoyment (*bhoga*).
Distinct from them, I am the witness,
pure consciousness, ever blissful.

"From me, all is born.
Through me, all is sustained.
In me, all is dissolved.
I am that Brahman, without a second.
Mayy eva sakalaṃ jātaṃ
mayi sarvaṃ pratishthitam
mayi sarvaṃ layaṃ yāti
tad brahmādvayam asmy aham

"I am smaller than small, yet also great.
I am this wonderful universe.
I am the ancient. I am Purusha.
I am the golden-hued Lord in the form of bliss.
Aṇoraṇīyān aham eva tadvan
mahān ahaṃ vishvam idaṃ vichitram
purātano 'haṃ purusho 'ham īsho
hiraṇmayo 'haṃ shiva-rūpam asmi

"Without hands and feet, I am of unfathomable power.
I see without eyes, hear without ears.
I alone know. Apart from my nature, there is no knower.
I am always the same—pure consciousness.
Apāṇi-pādo 'ham achintya-shaktiḥ
pashyāmy achakshuḥ sa shriṇomy akarṇaḥ
ahaṃ vijānāmi vivikta-rūpo
na chāsti vettā mama chit sadāham

"I am known indeed through all the Vedas.
I am the creator of Vedānta and the knower of Veda.
Good and evil do not touch me. There is no destruction
of me—having no birth, no body, no senses, no intellect.
Vedair anekair aham eva vedyo
vedānta-kṛid veda-vid eva chāham
na puṇya-pāpe mama nāsti nāsho
na janma dehendriya-buddhir asti

"In me there is no earth, water, fire,
air, or space. Thus having known
the nature of the highest Self, hidden in the heart,
without parts, without a second,
Na bhūmir āpo mama vahnir asti
na chānilo me 'sti na chāmbaraṃ cha
evaṃ viditvā paramātma-rūpaṃ
guhāshayaṃ nishkalam advitīyaṃ

"the witness of all, beyond what is and what is not,
one attains the pure nature of the highest Self.

Samasta-sākshiṃ sad-asad-vihīnaṃ
prayāti shuddhaṃ paramātma-rūpam

"Through this one obtains the knowledge
that destroys a flood of rebirths.
And thus by knowing this,
one attains the state of unity, *Kaivalya.*
Yes, one attains the state of unity."
Anena jñānam āpnoti
saṃsārārṇava-nāshanam
tasmād evaṃ viditvainaṃ
kaivalyaṃ padam ashnute
kaivalyaṃ padam ashnute

And so Āshvalāyana received the essence of all knowledge, Brahma Vidyā, from Lord Brahmā, having traveled to Brahma Loka.

In time Āshvalāyana composed *sūtras* on yagyas, known as the Āshvalāyana Gṛihya Sūtra and Āshvalāyana Shrauta Sūtra. He also became the head of a *gotra,* a Vedic family responsible for maintaining a branch, or *shākhā,* of the Veda.

अहं विश्वमिदं विचित्रम्

Ahaṃ vishvam idaṃ vichitram
I am this wonderful universe.
Kaivalya Upanishad, **20**

I am this wonderful universe

CHAPTER FOURTEEN

King Janaka Questions Yāgyavalkya

From the Bṛihadāraṇyaka Upanishad

In the city of Mithilā, the capital of the kingdom of Videha, lived the famous King Janaka. King Janaka enjoyed great wealth, and Videha was one of the most prosperous of the sixteen great kingdoms (*mahājanapada*) of ancient India. This was because King Janaka cultured spiritual values, which brought great benefit to himself and his kingdom. He was virtuous and kind to all, and deeply interested in philosophical questions.

Every morning King Janaka attended the *Agnihotra* offering, a yagya performed by the Brahmins for the prosperity of the kingdom. The Brahmins offered milk and ghee to the sacred fires as they chanted hymns from the Vedas. King Janaka was the *yajamāna,* or patron of the yagya, and so he sat on a special couch and witnessed the performance. Smoke from the fires rose through the yagya-shālā pavilion and the deep, rhythmical chanting of pandits created a serene feeling for all present.

One morning, the famous āchārya,Yāgyavalkya, sat in silence next to King Janaka at the Agnihotra offering. After the performance was over, Yāgyavalkya said, "Your

Majesty, I am very pleased with this Vedic yagya. I will grant you a boon. Ask for whatever you desire."

"This is the boon I wish for," King Janaka replied. "May the wise Yāgyavalkya answer any question that I might ask."

"Your wish is granted," said Yāgyavalkya.

With that, the king began to question the sage.

"My dear Yāgyavalkya," the king said, "what is the Self?"

"The Self is Ātmā," said Yāgyavalkya. "It is the light within the heart. The Self remains the same, even while a person thinks and moves about. When born, one obtains a body and becomes connected with this world. Upon departing, all is left behind.

"The Self is made of consciousness. It is made of mind. It is made of life. It is made of sight and hearing. It is made of earth, water, fire, air, and space. It is made of light and darkness together. It is made of desire and freedom from desire. It is made of anger and freedom from anger. It is this and it is that.

"The Self is incomprehensible, for it cannot be comprehended. It is indestructible, for it cannot be destroyed. It is the eye of the eye, the ear of the ear, and the mind of the mind.

"Whoever has found the Self, has awakened. He is the creator of the universe, the maker of all. The world is his. Indeed, he is the world."

"What is the experience of the Self?" asked King Janaka.

"The experience of the Self is the fulfillment of all desires," said Yāgyavalkya. "The spirit of man hastens to that place, like a falcon who soars in the sky and then folds his wings and flies home to his nest.

"When he thinks, 'I am everything' (*Aham evedaṃ sarvo 'smi*), then he is in the world of the spirit, free from desire, free from evil, free from fear.

"Like a child in the arms of his mother, he feels peace within. There is no sorrow. The spirit has crossed beyond good and evil, has crossed beyond the sorrows of the heart.

"There the spirit sees without seeing,
for there is nothing to see.
There is no second,
nothing other than himself to see.

"There the spirit speaks without speaking,
for there is nothing to say.
There is no second,
nothing other than himself with whom to speak.

"There the spirit hears without hearing,
for there is nothing to hear.
There is no second,
nothing other than himself to hear.

"There the spirit thinks without thoughts,
for there is nothing to think.
There is no second,
nothing other than himself about which to think.

"There the spirit knows without knowing,
for there is nothing to know.
There is no second,
nothing other than himself to know.

"This is for the person who experiences the Self."

"My dear Yāgyavalkya, could you speak about the world of dreams?" asked King Janaka.

"In the world of dreams there are no chariots and no roads, yet we create chariots and roads," replied Yāgyavalkya. "There are no pleasures and delights, yet we create pleasures and delights. There are no ponds, lotus pools, and rivers, yet we create ponds, lotus pools, and rivers.

"The world of dreams is between this world and the next. In the world of dreams the Self reflects upon the joys of this world and also sees the joys of the world to come.

"There is a verse that says:

"Abandoning his physical form in sleep,
the wakeful Self witnesses his sleeping body.
By his own light he returns home,
the spirit of golden radiance, the swan of purity.
Svapnena shārīram abhiprahatyāsuptaḥ
suptān abhichākashīti
shukram ādāya punar aiti sthānaṃ
hiraṇmayaḥ purusha eka-haṃsaḥ

"As his vital breath guards his nest below,
the immortal spirit soars afar.

He flies wherever he desires,
the spirit of golden radiance, the swan of purity.
Prāṇena rakshann avaraṃ kulāyaṃ
bahish kulāyād amṛitash charitvā
sa īyate amṛito yatra kāmaṃ
hiraṇmayaḥ purusha eka-haṃsaḥ

"He goes above and below, up and down,
in his state of dreams. He becomes like a god
who creates many forms for himself, now enjoying
beauty and laughter, now seeing fearful sights.
Svapnānta uchchāvacham īyamāno
rūpāṇi devaḥ kurute bahūni
uteva strībhiḥ saha modamānaḥ
jakshat utevāpi bhayāni pashyan

"When the spirit returns from the land of dreams," Yāgyavalkya continued, "where he has seen good and bad, whatever he has seen does not return with him, for he is not attached. As a fish swims between two banks of a river, so the Self moves between the world of waking and the world of dreaming."

"I give you a thousand cows," said King Janaka. "Please tell me, what is real joy?"

"Joy is Brahman," said Yāgyavalkya. "Most human beings enjoy only a small portion of the joy of Brahman.

"Your Majesty, imagine a person who is healthy, wealthy, admired by others, and provided with all human pleasures. This is the greatest joy for humans.

"A hundred times this joy
is the joy of the world of the ancestors.
A hundred times the joy of the ancestors
is the joy of the world of the Gandharvas.
A hundred times the joy of the Gandharvas
is the joy of the world of the Devas.
A hundred times the joy of the Devas
is the joy of the world of Prajāpati.
A hundred times the joy of Prajāpati
is the joy of Brahman.

"This is the joy of one who is pure and free of desire. This is the supreme joy, the highest bliss, the world of the spirit."

"I give you a thousand cows," said King Janaka. "Please tell me, O Yāgyavalkya, what happens at the time of death."

"Your Majesty," said Yāgyavalkya, "think of how a cart, loaded with heavy stones, groans and creaks when it moves. Like that, the body is a cart, and it groans as a man approaches death and is breathing with difficulty.

"Think of how a fig is loosened from its stem. Like that, when the body becomes thin and weak from old age or disease, the spirit in man releases itself from the body. Then the spirit returns by the same way to the place from which it started.

"Think of when a king approaches a village. The nobles and officers, the charioteers and the heads of the village prepare food, drink, and lodging. They say, 'The king is coming! Here he comes!' Like that, all the powers of nature wait for him who is approaching death, saying, 'Brahman is coming! Here he comes!'

"Think of when a king is about to depart from the village. The nobles and officers, the charioteers and the heads of the village gather around him and bid him farewell. Like that, when a man is giving up the breath of life, all the powers of nature gather 'round.

"When the person becomes weak and seems unconscious, all the powers of nature gather 'round; then the person collects his vital powers and enters into his own heart.

" 'He is becoming one; he sees no more,' they say.
'He is becoming one; he speaks no more,' they say.
'He is becoming one; he hears no more,' they say.
'He is becoming one; he thinks no more,' they say.
'He is becoming one; he knows no more,' they say.

"Then a light shines in his heart, and this light guides him as he departs. The vital breath departs with him, and the other senses go with the vital breath. As he enters into life's unbounded intelligence, his own intelligence departs with him—his ageless wisdom, his deeds, and his past experience. They as if take him by the hand and guide him.

"Your Majesty, think of a caterpillar and how it climbs to the tip of a blade of grass. See how it reaches out to another blade of grass and draws itself over to it. Like that, the spirit departs from this body and its ignorance, and reaches out to another body and draws itself to it.

"Think of how a goldsmith takes an old piece of gold and creates a new and more beautiful form. Like that, the spirit, having departed from the body and its ignorance, creates another new and more beautiful form, like the form of his ancestors, or a Gandharva, or a Deva, or other beings."

"I give you a thousand cows," said King Janaka. "O Yāgyavalkya, what of desires that remain at the time of death?"

Yāgyavalkya answered, "The mind goes towards the object it desires. Of this it is said,

"If he desires the world of fathers,
by his mere intention (*saṃkalpa*), fathers arise.
Living in the world of fathers, he is filled with joy.
Sa yadi pitṛi-loka-kāmo bhavati
saṃkalpād evāsya pitaraḥ samuttishthanti
tena pitṛi-lokena sampanno mahīyate

"If he desires the world of mothers,
by his mere intention, mothers arise.
Living in the world of mothers, he is filled with joy.
Atha yadi mātṛi-loka-kāmo bhavati
saṃkalpād evāsya mātaraḥ samuttishthanti
tena mātṛi-lokena sampanno mahīyate

"If he desires the world of friends,
by his mere intention, friends arise.
Living in the world of friends, he is filled with joy.

Atha yadi sakhi-loka-kāmo bhavati
saṃkalpād evāsya sakhāyaḥ samuttishthanti
tena sakhi-lokena sampanno mahīyate

"If he desires the world of perfumes and garlands,
by his mere intention, perfumes and garlands arise.
Living in the world of perfumes and garlands,
he is filled with joy.
Atha yadi gandha-mālya-loka-kāmo bhavati
saṃkalpād evāsya gandha-mālye samuttishthanti
tena gandha-mālya-lokena sampanno mahīyate

"If he desires the world of song and music,
by his mere intention, song and music arise.
Living in the world of song and music,
he is filled with joy.'
Atha yadi gīta-vādita-loka-kāmo bhavati
saṃkalpād evāsya gīta-vādite samuttishthanti
tena gīta-vādita-lokena sampanno mahīyate

"He does not, however, stay in those worlds forever," Yāgyavalkya continued. "When the fruits of his actions in this world are exhausted, then he comes again from that world to this world. He who has desires is born again by virtue of those desires. He is born again along with his desires. This is for the person who desires."

"I give you a thousand cows," said King Janaka. "What of freedom from desire?" he asked.

"Freedom from desire comes from knowing the Self," Yāgyavalkya replied. "If a person knows the Self, he knows 'I am this.' Then what can he wish for? He has fulfilled his desires. What desire takes him to another body? Those whose desires are satisfied are fulfilled souls—all their desires vanish even here on earth.

"There is a verse that says:

"When all the desires that dwell deep
in the heart are cast away,
then a mortal becomes immortal,
then he attains Brahman.
Yadā sarve pramuchyante
kāma ye 'sya hṛidi shritāḥ
atha martyo 'mṛito bhavati
atra brahma samashnute

"While we live in this life," Yāgyavalkya continued, "we may become fulfilled, our desires satisfied. And if we do not, then the darkness is deep. Those who have fulfilled their desires in the Self become immortal. They are the knowers of Brahman.

"For those who have freedom from desire, at the final hour, when they think of themselves, they understand this truth about themselves:

"You are indestructible.
You are unshaken.
You are the essence of life."

Akshitam asi
achyutam asi
prāṇa-saṃshitam asi

Then King Janaka asked his final question, "What is Brahman?"

"Brahman is infinite, eternal," said Yāgyavalkya. "It is beyond space, unborn, the great, the stable. When one clearly beholds Brahman then one sees the ruler of what was and what will be. Then one becomes free from fear.

"The knower of Brahman has no bonds of attachment, for he is free. He is beyond suffering. He is beyond fear. Evil does not overcome him; he overcomes evil.

"The knower of Brahman is not overwhelmed by exultation or grief from his past actions, whether good or bad. He is beyond both. What he has done and what he has left undone do not affect him. He does not ponder many words, for many words are weariness.

"The knower of Brahman has found peace. He is calm. He is patient. He is composed, free from doubt. He has found the treasure of life, the great spirit of the universe, and so he is tranquil.

"There is a verse which says,

"I have found a small path, ancient,
which stretches far away.
On this path, the wise, the knowers of Brahman,
rise to the heavenly worlds and are liberated.

Aṇuḥ panthā vitataḥ purāṇaḥ
māṃ spṛishto 'nuvitto mayaiva
tena dhīrā api yanti brahmavidaḥ
svargaṃ lokam ita ūrdhvaṃ vimuktāḥ

"That path, they say, is white, blue, yellow,
green, and red. That path to Brahman
is traversed by the knowers of Brahman,
who perform right action and are filled with light."
Tasmin shuklam uta nīlam
āhuḥ pingalaṃ haritaṃ lohitaṃ cha
esha panthā brahmaṇā hānuvittaḥ
tenaiti brahmavit puṇyakṛit taijasash cha

"I give you the kingdom of Videha," King Janaka said, "and I give you myself as your servant."

And so in the kingdom of Videha the dialogue between the great Yāgyavalkya and the famous King Janaka came to an end. Teacher and student, Ṛishi and Rājā, they shared together the highest knowledge, the knowledge of the Self. Through this timeless wisdom King Janaka gained eternal liberation in enlightenment and created a prosperous kingdom where people experienced the Self and so enjoyed harmony and happiness in their lives.

असतो मा सद्गमय
तमसो मा ज्योतिर्गमय
मृत्योर्मा अमृतं गमय

Asato mā sad gamaya
tamaso mā jyotir gamaya
mṛityor mā amṛitaṃ gamaya

From non-existence lead us to existence,
From darkness lead us to light,
From death lead us to immortality.
Bṛihadāraṇyaka Upanishad, 1.3.28

From non-existence lead us to existence
From darkness lead us to light
From death lead us to immortality

CHAPTER FIFTEEN

The Moving and the Unmoving

From the Īsha Upanishad

Once, in ancient India, there was an āchārya who taught lessons by the side of the Ganges, the most sacred river of India. Under a large banyan tree next to the holy river, called the River of Heaven, the āchārya sat early every morning.

Beyond the river rose the wooded hillsides of the high Himālayas. The tall blue mountains trapped the billowing clouds and caused them to release their rain. Hidden in the high mountain valleys were holy shrines, such as Kedārnāth and Badrīnāth, which attracted devout pilgrims.

Early one morning, as the sun still hid behind the tall mountains, the āchārya was teaching his young students. His message was a simple one: life is naturally full. His students, seated on the grass around him, were learning the highest wisdom of all—that one who lives fullness fathoms the silence within and also enjoys the world outside.

The āchārya began the morning's lesson:

"That is full; this is full.
From fullness, fullness comes out.
Taking fullness from fullness,
what remains is fullness.

Pūrṇam adaḥ pūrṇam idaṃ
pūrṇāt pūrṇam udachyate
pūrṇasya pūrṇam ādāya
pūrṇam evāvashishyate
Oṃ shāntiḥ shāntiḥ shāntiḥ

"Now, dear students, do you have any questions?" asked the āchārya.

"Honored Sir, would you speak about the Self?" asked one student.

The āchārya responded:

"Everything here, whatever lives in the moving world,
dwells within the unmoving Lord.
Enjoy it by going beyond.
Do not seek what belongs to others.
Īshāvāsyam idaṃ sarvaṃ
yat kiṃ cha jagatyāṃ jagat
tena tyaktena bhunjīthā
mā gṛidhaḥ kasya svid dhanam

"Thus performing actions here,
one may aspire to live a hundred years.
For a man there is no other way
whereby your actions will not bind.
Kurvann eveha karmāṇi
jijīvishech chhataṃ samāḥ
evaṃ tvayi nānyatheto 'sti
na karma lipyate nare

"There are indeed worlds without light,
covered in blinding darkness.
There, after departing,
go those who slay the Self.
Asūryā nāma te lokā
andhena tamasā 'vṛitāḥ
tāṃs te pretyābhigachchhanti
ye ke chātmahano janāḥ

"The Self is unmoving, one, yet swifter than the mind.
The senses cannot fathom it—it is ever beyond their grasp.
Standing still, it overtakes those who run.
Yet in it rests the breath of all that move.
Anejad ekaṃ manaso javīyo
nainad devā āpnuvan pūrva-marshat
tad dhāvato 'nyān atyeti tishthat
tasminn apo mātarishvā dadhāti

"It moves, yet it moves not.
It is far, yet it is near.
It is within all this,
and yet outside all this.
Tad ejati tan naijati
tad dūre tad vantike
tad antarasya sarvasya
tad u sarvasyāsya bāhyataḥ

"He who sees
all beings in the Self
and the Self in all beings—
such a seer withdraws from nothing.
Yas tu sarvāṇi bhūtāni
ātmany evānupashyati
sarva-bhūteshu chātmānaṃ
tato na vijugupsate

"For the wise, all beings have become the Self.
What delusion, then,
what sorrow can befall him
who has perceived the oneness?
Yasmin sarvāṇi bhūtāni
ātmaivābhūd vijānataḥ
tatra ko mohaḥ kaḥ shoka
ekatvam anupashyataḥ

"He is all-pervading, radiant, formless, invincible,
pure, unblemished, free of imperfections.
He is the seer, the knower of all,
the encompasser of all, the self-existent.
Throughout endless time he has created
innumerable objects in due succession."
Sa paryagāch chhukram akāyam avraṇam
asnāviraṃ shuddham apāpaviddham
kavir manīshī paribhūḥ svayambhūr
yāthātathyato 'rthān vyadadhāch
chhāshvatībhyaḥ samābhyaḥ

The students seemed enraptured with the discourse of the āchārya. However, the āchārya could see that several of them were puzzled. And so he said,

"My dear students, would any of you like to ask a question?"

Eager to understand, one of them asked, "Sir, could you tell us about knowledge and ignorance?"

The āchārya looked at his beloved students, and decided they were ready to learn about knowledge and ignorance:

"Into blinding darkness go
those who are devoted to ignorance.
And into even greater darkness
enter those who delight in knowledge."
Andhaṃ tamaḥ pravishanti
ye 'vidyāṃ upāsate
tato bhūya iva te
tamo ya u vidyāyāṃ ratāḥ

This startled some of the students, for they wanted to be devoted to knowledge and nothing else. "What can our teacher mean? Is not knowledge the highest reality of life?" they thought.

As if responding to their thoughts, the āchārya continued,

" 'It is other than knowledge,' they say.
'It is other than ignorance,' they say.
Thus have we heard from the wise
who have explained this to us.

Anyad evāhur vidyayā
'nyad āhur avidyayā
iti shushruma dhīrāṇāṃ
ye nas tad vichachakshire

"Knowledge and ignorance—
he who knows both together,
through ignorance goes beyond death,
and through knowledge attains immortality."
Vidyāṃ chāvidhyāṃ cha
yas tad vedobhayaṃ saha
avidyayā mṛityuṃ tīrtvā
vidyayā 'mṛitam ashnute

Some of the students understood immediately and were delighted. They heard the teacher say, "he who knows both together," and they knew that he spoke of the highest reality, Brahman, which accepts everything. That is why it is called the totality.

Other students, however, felt lost. "Sir," one boy said, "could you give us another example?"

"Yes, my dear," said the āchārya.

"Into blinding darkness go
those who are devoted to the unmanifest.
And into even greater darkness
enter those who delight in the manifest."
Andhaṃ tamaḥ pravishanti
ye 'sambhūtim upāsate

tato bhūya iva te
tamo ya u sambhūtyāṃ ratāḥ

Now more of the students were starting to understand. They realized that Brahman includes both the manifest and the unmanifest.

The āchārya continued,

" 'It is other than the manifest,' they say.
'It is other than the unmanifest,' they say.
Thus have we heard from the wise
who have explained this to us.
Anyad evāhuḥ sambhavād
anyad āhur asambhavāt
iti shushruma dhīrāṇāṃ
ye nas tad vichachakshire

"Manifest and unmanifest—
he who knows both together,
through the unmanifest goes beyond death,
and through the manifest attains immortality."
Sambhūtiṃ cha vināshaṃ cha
yas tad vedobhayaṃ saha
vināshena mṛityuṃ tīrtvā
sambhūtyā 'mṛitam ashnute

A wave of satisfaction rippled through the air. All the students grasped the profound meaning, the true nature of Brahman as wholeness.

As if to offer a blessing to the students, just then the first rays of the sun peeked over the lofty mountains, sending warmth and light, and causing the birds to burst into song. As the ācharya looked at the sparkling reflection of the morning sun on the Ganges, he knew that he had spoken the truth, and that this truth was hidden only by the thinnest of veils. Then he offered a short invocation, wishing that all may clearly perceive the immortal life. Quietly he whispered,

"The face of truth
is hidden by a disc of gold.
O Pūshan, unveil it so that I,
who love the truth, may see it.
Hiraṇmayena pātreṇa
satyasyāpihitaṃ mukham
tat tvaṃ pūshann apāvṛiṇu
satya-dharmāya dṛishtaye

"O Pūshan, the lone traveler,
O Yama, O sun, the giver of life,
withdraw your rays and gather in your light,
so I may behold your most beautiful form!
Whoever is that person—I am he.
Pūṣhann ekarshe yama sūrya prājāpatya
vyūha rashmīn samūha tejaḥ
yat te rūpaṃ kalyāṇatamaṃ tat te
pashyāmi yo 'sāv asau purushaḥ so 'ham asmi

"May this life join with the immortal breath.
Then may this body end in ashes.
Remember, mind, remember what has been done!
Remember, mind, what has been done, remember!
Vāyur anilam amṛitam
athedaṃ bhasmāntaṃ sharīram
oṃ krato smara kṛitaṃ smara
krato smara kṛitaṃ smara

"O Agni, guide us along the virtuous path to prosperity.
You, Lord, who knows all our deeds,
remove from us our wrongdoings.
We offer you bountiful words of praise."
Agne naya supathā rāye asmān
vishvāni deva vayunāni vidvān
yuyodhyasmaj juhurāṇam eno
bhūyishthāṃ te nama-uktiṃ vidhema

Then the āchārya finished the lesson with the same words with which he had begun,

"That is full; this is full.
From fullness, fullness comes out.
Taking fullness from fullness,
what remains is fullness."
Pūrṇam adaḥ pūrṇam idaṃ
pūrṇāt pūrṇam udachyate
pūrṇasya pūrṇam ādāya
pūrṇam evāvashishyate
Oṃ shāntiḥ shāntiḥ shāntiḥ

With this the lesson was finished. The students thanked their beloved teacher for this precious wisdom. One by one they bowed and touched his feet.

Far in the distance the Himālayas stood unmoving, their white peaks towering above the clouds. Flowing from the mountains, the sacred Ganges glimmered in the morning sun like the matted locks of Lord Shiva. The unmoving mountains and the moving river—these two are one. The moving and the unmoving are one. This is the supreme wisdom of life, which the āchārya gave to his students by the side of the Ganges, the River of Heaven.

पूर्णमदः पूर्णमिदं पूर्णात्पूर्णमुदच्यते
पूर्णस्य पूर्णमादाय पूर्णमेवावशिष्यते

Pūrṇam adaḥ pūrṇam idaṃ pūrṇāt pūrṇam udachyate
pūrṇasya pūrṇam ādāya pūrṇam evāvashishyate

That is full; this is full. From fullness, fullness comes out.
Taking fullness from fullness, what remains is fullness.

Isha Upanishad, Shānti Pātha

All good I should hear from the ears
All good I should see through the eyes

CHAPTER SIXTEEN

Garland of Questions

From the Prashna Upanishad

A group of friends named Kabandhī, Bhārgava, Kausalya, Gārgya, Satyakāma, and Sukesha treated each other with kindness and shared a common goal—each fervently wished to understand Brahman. They were devoted to Brahman, intent on Brahman, and in search of Brahman.

After some time, they realized that they could not fathom the subtleties of Brahman by themselves. They knew that they needed the guidance of an enlightened teacher. So one morning they left their small village and set out on foot for the āshram of Pippalāda, a famous teacher who was given that name due to his fondness for the *pippala* fruit.

For several days they walked until they reached a tall, cool forest, called Dashāraṇya. It was here that Pippalāda had meditated for a long time, motionless as a mountain. It was said that he created such an atmosphere of peace that every animal, even the white tiger, lived in harmony with every other animal. From his meditations, Pippalāda was granted a boon, called *sarva-kāma-siddhi,* which gave him anything he desired.

As they walked through the forest one afternoon, the six friends heard the sound of pandits reciting the Vedas and knew that Pippalāda's āshram must be near. A *brahmachārī*, one who has devoted his life to the study of Vedic knowledge, greeted them at the entrance to the āshram. He bathed their feet in water, rubbed cool sandalwood paste on their foreheads, and asked them to rest in the shade, where he served them food and water.

As the silence of the āshram settled on them in the cool evening air, the six friends were led inside a small cottage to meet Ṛishi Pippalāda. They smelled incense and saw him sitting on a simple cot, his legs folded. The friends bowed respectfully, and Ṛishi Pippalāda motioned for them to sit.

"Sir," Kabandhī asked, "would you be so kind as to answer our questions about Brahman?"

"Yes, I will answer your questions," Ṛishi Pippalāda quietly said. "But first I would like you to live in the āshram for one year."

The six friends were surprised at the request of the teacher. Why did they need to wait for a year? They were expecting Pippalāda to tell them about Brahman, as one tells another about a flowering tree or a colorful bird. What would he tell them in a year that he could not tell them now?

However, they humbly said, "Yes, honored Sir," touched the teacher's feet in respect, and quietly left the small cottage.

That evening the six friends were given sleeping mats and were led to a small hut to sleep in. The next morning, like the other pupils, they awoke before dawn, at the song of the woodcock. This was the time called *brahma-muhūrta,* the hour of Brahmā. After bathing in the river and practicing yoga postures, the young men gathered under the large *nyagrodha* tree in the courtyard. There they sat, eyes closed, facing the rising sun.

As time went by, their morning and evening meditations became deeper. Gradually they felt more and more vital, revived in body and spirit. Gradually they felt inner happiness unfolding like the petals of a lotus. Ṛishi Pippalāda watched their progress and was pleased.

A year passed by. The large nyagrodha tree in the courtyard looked just as it had the year before. However, the six friends were completely transformed. Their minds and hearts were filled with appreciation for everything around them.

One morning as they sat after meditation, a brahmachārī asked them to come to Ṛishi Pippalāda's cottage. As they sat at the feet of their teacher, they noticed how the sun's morning rays filtered through the shutters, bathing the room in a soft glow. They noticed that Ṛishi Pippalāda was radiant with the inner light of the Self. They felt gratitude to be in the presence of such an enlightened soul and in such a heavenly atmosphere. And they wondered, "How could we have missed this a year ago?"

Ṛishi Pippalāda greeted them and began by chanting a traditional invocation:

"All good I should hear from the ears.
All good I should see through the eyes.

Bhadraṃ karṇebhiḥ shṛiṇuyāma devāḥ
bhadraṃ pashyemākshabhir yajatrāḥ

"May we, with strong limbs, enjoy
the long life given to us by nature.
Sthirair angais tushtuvāṃsas tanūbhiḥ
vyashema devahitaṃ yadāyuḥ

"May Indra, filled with glory, grant us well-being.
May Pūshan, the knower of all, grant us well-being.
Svasti na indro vṛiddha shravāḥ
svasti naḥ pūshā vishva-vedāḥ

"May Tārkshya, whose path is safe, grant us well-being.
May Bṛihaspati grant us well-being."
Svasti nas tārkshyo arishta-nemiḥ
svasti no bṛihaspatir dadhātu

The six friends listened to the invocation and were absorbed in quiet happiness. Ṛishi Pippalāda smiled. He knew that now their minds were ready to understand the subtleties of Brahman.

"Do you have any questions, my dear students?" Ṛishi Pippalāda gently asked.

"Venerable Sir, I indeed have a question," Kabandhī said. "Could you tell us, Sir, from where all beings are born?"

"The lord of creation desired to create beings," Ṛishi Pippalāda said, "and so he retired to meditate for a thousand years. After his meditations, he created spirit and matter. These two were like opposite twins—one was without form, and the other was with form. Spirit, though without form, breathed life into matter.

"From these two came the sun (*Āditya*) and the moon (*Chandramā*). The sun, rising in the east, is the giver of heat. Having a thousand rays, it warms everything, including the moon.

"Like the sun and moon, and the day and night, many opposites arise in creation, and the crown of creation is the human being. Through days and nights on earth, human beings traverse the path of truth until they realize Brahman, thus satisfying the first desire of the Lord of Creation."

"Thank you, Sir," said Kabandhī, who was completely satisfied with Ṛishi Pippalāda's answer.

"Is there another question?" Ṛishi Pippalāda gently asked.

"Venerable Sir, I indeed wish to ask a question," Bhārgava said. "Could you tell us what powers sustain and support the body, and which is the greatest of them?"

Ṛishi Pippalāda told them a story. "Speech (*vāk*), mind (*manas)*, sight (*chakshu*), and hearing (*shrotra*) claimed to sustain the body. They said, 'We sustain and support the body.' However, the vital breath (*prāṇa*) heard them and said, 'I alone sustain and support the body.'

"The other powers replied, 'No! We support and sustain the body!' And so prāṇa decided to teach them a lesson. Prāṇa went straight up—out of the body. Then the others were forced to go straight up. Then prāṇa sprang down, back into the body. And the others gratefully followed. Now they believed him.

"And so, you see, prāṇa is like the queen bee. When the queen bee goes up, the other bees follow. When the queen bee goes down, the other bees follow. Prāṇa protects the body as a loving mother protects her children. While there are many powers that sustain and support the body, the greatest of them is prāṇa."

With that, Bhārgava was completely satisfied.

"Is there another question?" Ṛishi Pippalāda gently asked.

"Venerable Sir, I indeed wish to ask a question," Kausalya said. "What are the divisions of prāṇa? What is the basis of prāṇa?"

"Prāṇa is the vital breath, located in the head and chest," Pippalāda said. "Besides prāṇa, there is the upward breath, *udāna,* located in the chest and throat; the even breath, *samāna,* responsible for digestion; the downward breath, *apāna,* responsible for elimination; and the moving breath, *vyāna,* responsible for circulation.

"The basis of prāṇa is Ātmā, the Self. The Self is located in the heart. Prāṇa follows the Self as a shadow follows a person. Through prāṇa the Self creates the world."

With that, Kausalya was completely satisfied.

"Is there another question?" Ṛishi Pippalāda gently asked.

"Venerable Sir, I indeed wish to ask a question," Gārgya said. "Could you tell us who it is that sleeps, who it is that dreams, who it is that rests in the Self?"

"When the sun sets," Pippalāda said, "all the rays of light gather together and become one with the sun. When the sun rises, the rays again shoot forth. Like this, when all experiences draw together in the mind, then the person does not see, does not hear, and does not speak. The person sleeps. As the person sleeps, the vital breath conducts the activities of the body.

"Sometimes the person sees again what was seen before, hears again what was heard before, and experiences again what was experienced before. Then the person dreams.

"At other times, the person is overcome by a spiritual brilliance (*tejas*), and happiness rises up in the body. In this state, the person sees no objects. Just as birds rest in a tree, so the person rests in the Self. Then the person experiences *turīya*, called the fourth. This is the experience of the Self.

"The Self is the seer, the hearer, the knower, the doer. The Self is Purusha. The Self is pure, without form, and without color. The Self is supreme. It is imperishable. It is beyond all, and so is called the transcendental Self (*Paramātmā*). It is all-knowing, the foundation of all, the one light, the life of creation.

"Those who realize the Self know everything and become everything. They know the whole world."

With that, Gārgya was completely satisfied.

"Is there another question?" Ṛishi Pippalāda gently asked.

"Venerable Sir, I indeed wish to ask a question," Satyakāma said. "How does one know the Self?"

"The Self is known through meditation," Pippalāda said. "The lower Brahman and the higher Brahman are both known through meditation.

"Through meditation one experiences greatness in the world of human beings. This is the lower Brahman, called *apara* Brahman.

"Through meditation one experiences evenness, *sāma*. Established in evenness one knows with clarity the Self, the inner splendor of light. This is *para* Brahman, the higher Brahman. Experiencing this, a person is released from past deeds as a serpent is released from its old skin. He attains the tranquility of the Self, beyond old age and death, beyond fear. He attains the supreme."

With that, Satyakāma was completely satisfied.

"Is there another question?" Ṛishi Pippalāda gently asked.

"Venerable Sir, I indeed wish to ask a question," Sukesha said. "Who is Purusha?"

"Purusha resides within the body," Pippalāda said. "Like spokes on a wheel, all parts of the body are fixed upon Purusha, the Self. Like rivers merging in the ocean, these names and forms disappear upon the experience of Purusha.

"From Purusha arise all the oceans and mountains. From Purusha arise the herbs and their healing sap. Purusha, indeed, is this whole world. He is Brahman, the supreme, the immortal. He who knows Purusha, hidden in the heart, cuts the knot of ignorance here on earth.

"Of Purusha it is said,

"Purusha has a thousand heads,
a thousand eyes, a thousand feet.
Having encompassed the earth on all sides,
he extends beyond by ten fingers' width.

"With hands and feet everywhere,
eyes, heads, and faces everywhere,
with ears everywhere, he stands,
embodying all in the world.

"Purusha is truly all this,
what has been and what will be.
He is the Lord of immortality,
who dwells in the heart of every being."

With that, Sukesha was completely satisfied.

"Is there another question?" Ṛishi Pippalāda gently asked.

There was silence, since there were no more questions.

"My beloved students," said Pippalāda, "you have given me a beautiful garland—a garland of questions. Take this precious wisdom to your families and help them to know Brahman, for Brahman is the goal of life. There is nothing higher."

With this Pippalāda concluded his lesson. The six friends thanked him with all their hearts. Then they left the āshram and returned home, having obtained the totality, Brahman. And throughout their lives they remained true friends, because they had helped each other unfold the purpose of their lives, the realization of Brahman.

भद्रं कर्णेभिः शृणुयाम देवाः
भद्रं पश्येमाक्षभिर्यजत्राः

Bhadraṃ karṇebhiḥ shṛiṇuyāma devāḥ
bhadraṃ pashyemākshabhir yajatrāḥ

All good I should hear from the ears.
All good I should see through the eyes.

Prashna Upanishad, Shānti Pātha

CHAPTER SEVENTEEN

Satyakāma Teaches Upakosala

From the Chhāndogya Upanishad

Satyakāma was a teacher who lived in a tranquil āshram beside the forest. Many years before, when he was young, Satyakāma had learned Brahma Vidyā, the supreme knowledge, from his teacher, Gautama. Before giving Satyakāma the teaching, Gautama made an unusual request. He asked Satyakāma to tend four hundred cows in the forest and to return when they had multiplied to one thousand.

Satyakāma lived in the forest for many years, meditating and tending the cows, and naturally he began to experience a deep inner happiness. In time the cows multiplied to a thousand, and he started back to Gautama's āshram. On the way back, a bull, a swan, and a fire offered Satyakāma the wisdom of Brahman. When Gautama saw Satyakāma's illuminated face, he said, "I see that you have found Brahman!" Then Gautama gave Satyakāma the final teaching, Brahma Vidyā.

Years later, when Satyakāma had grown older, he taught his own students how to realize Brahman. He had his own forest āshram, where cows, deer, and wild animals wandered freely, living together in harmony. Each morning his young students gathered around him for lessons, just as Satyakāma had learned from his teacher when he was a boy.

Satyakāma taught his beloved students everything that he had learned. The students learned how to perform the various yagyas. There were many duties in the performance of a yagya, and the students learned them all.

They learned how to build the yagya-shālā, the cedar hall for the yagya, and where to place the fires in the hall. In the center was the elevated ground, called the *vedi*. To the west of the vedi was the householder's fire, called the western fire (*gārhapatya*). To the east of the vedi was the eastern fire (*āhavanīya*), and to the south was the southern fire (*dakshiṇa*).

The students learned how to prepare the ground, how to build the altar with bricks, how to prepare the offerings of ghee and puffed rice, and how to make the offerings to the various fires used for the yagya.

They learned to chant with the traditional melody—*udātta* for the middle tone, *anudātta* for the low tone, and *svarita* for the high tone. They learned *gāyatrī* and the other traditional meters.

Satyakāma taught them to become established in the Self, and then to chant the slow, rhythmical melodies of Sāma Veda. He said, "By singing Sāma Veda, you will obtain immortality from the Devas, the powers of nature."

One of Satyakāma's favorite students was Upakosala, the son of Kamala. Upakosala had studied in the āshram for twelve years, which was the tradition. At the end of their twelve years of study, all the other students returned to their homes. But Satyakāma asked Upakosala to remain in the āshram after the others had left.

Satyakāma's wife noticed what had happened, and she said to Satyakāma, "My dear Satyakāma, Upakosala has been your devoted student all these twelve years. He has performed the yagyas with honor, three times each day. Please give him the final teaching so that he may return home."

But Satyakāma did not give Upakosala the final instruction. Instead, he went away on a journey.

Then Upakosala became sick and was unable to eat. "Dear Upakosala, why have you stopped eating your food?" asked Satyakāma's wife in concern.

"I have many desires in many directions," he said. "Now I am filled with grief; therefore I cannot eat."

Satyakāma's wife did not know what to do. But then something miraculous happened. The three yagya fires, which had been tended so lovingly by Upakosala and the other students, began to talk among themselves! They said, "Upakosala has meditated here in the āshram for twelve years. He has served his teacher and has learned to perform the various yagyas with honor. Why don't we give him the final teaching?"

The fires began to speak. "Dear Upakosala," they said, "life is Brahman. Happiness is Brahman. Space is Brahman."

Then the gārhapatya, the western fire, spoke to him. "O Upakosala, do you know the person who resides in the sun (*Āditya*)?"

"Yes," said Upakosala, "I am familiar with the person who resides in the sun."

The western fire said,

"The person who is seen in the sun,
I am he. Indeed I am he!
Ya esha āditye purusho dṛishyate
so 'ham asmi sa evāham asmi

"With this knowledge, O Upakosala, you will be free from the unfavorable influence of past action, you will enjoy this world, you will live a long and full life, and your descendants will live in abundance."

Then the dakshiṇa, the southern fire, spoke to him, "O Upakosala, do you know the person who resides in the moon (*Chandramā*)?"

"Yes," said Upakosala, "I am familiar with the person who resides in the moon."

The southern fire said,

"The person who is seen in the moon,
I am he. Indeed I am he!
Ya esha chandramasi purusho dṛishyate
so 'ham asmi sa evāham asmi

"With this knowledge, O Upakosala, you will be free from the unfavorable influence of past action, you will enjoy this world, you will live a long and full life, and your descendants will live in abundance."

Then the āhavanīya, the eastern fire, said to him, "O Upakosala, do you know the person who resides in lightning (*Vidyut*)?"

"Yes," said Upakosala, "I am familiar with the person who resides in lightning."

The eastern fire said,

"The person who is seen in lightning,
I am he. Indeed I am he!
Ya esha vidyuti purusho dṛishyate
so 'ham asmi sa ẹvāham asmi

"With this knowledge, O Upakosala, you will be free from the unfavorable influence of past action, you will enjoy this world, you will live a long and full life, and your descendants will live in abundance."

Upakosala reflected upon what the fires had said, and he thought, "The Self is in everything."

Then the fires spoke together. "Dear Upakosala, now you have the knowledge of the Self, *Ātma Vidyā*. When your teacher returns, he will give you the final teaching of Brahman."

In a few days Satyakāma returned from his journey. He saw the radiant smile on Upakosala's face. "My dear Upakosala," he said, "I see that you have found Brahman!" Satyakāma remembered when his own teacher, Gautama, had spoken these same words to him, so long ago, and now Satyakāma was pleased to see his own student looking so brilliant.

"My dear Upakosala," asked Satyakāma, "who has instructed you?"

"Sir, they have instructed me," Upakosala said as he pointed to the fires.

Satyakāma remembered when the fire had also instructed him, when he lived as a youth in the forest. He said, "Now, dear Upakosala, I will give you the final teaching. Just as water does not cling to the lotus leaf, unfavorable karma will not cling to one who knows this knowledge. Upakosala, do you know the person who resides in the eye?"

"Yes," said Upakosala, "I am familiar with the person who resides in the eye."

Satyakāma said,

"The person who is seen in the eye is Ātmā.
This is the immortal, beyond fear. This is Brahman."
Ya esho 'kshiṇi purusho dṛishyata esha ātmā
etad amṛitam abhayam etad Brahma

Upakosala reflected upon what Satyakāma had said, and he thought, "The Self in everything is also in me. Therefore, when I know myself, I know everything."

Then Satyakāma said,

"The knower of Ātmā attracts everything beautiful.
The knower of Ātmā is surrounded by everything beautiful.
The knower of Ātmā radiates in all worlds.
The knower of Ātmā attains Brahman.
Yes, the knower of Ātmā attains Brahman."

And then Satyakāma told Upakosala that he could go home, now that he had received the final teaching, Brahma Vidyā. Filled with joy, Upakosala returned home. Many years later, he himself became a teacher and taught his own students to recite the Veda and to perform the yagyas. He gave them the final teaching of Brahma Vidyā, just as Satyakāma had given it to him, and Gautama had given it to Satyakāma. And this is how the supreme teaching has been passed, from teacher to student, in an uninterrupted tradition since time immemorial in Veda Bhūmi, the land of the Veda, the land of knowledge.

य एषोऽक्षिणि पुरुषो दृश्यत एष आत्मा
एतदमृतमभयमेतद् ब्रह्म

Ya esho 'kshiṇi purusho dṛishyata esha Ātmā
etad amṛitam abhayam etad Brahma

The person who is seen in the eye is Ātmā.
This is the immortal, beyond fear. This is Brahman.
Chhāndogya Upanishad, 4.15.1

CHAPTER EIGHTEEN

Pratardana Learns about Wholeness

From the Kaushītaki Upanishad

Once there was a young prince (*rājaputra*) named Pratardana, who was the son of the famous King Divodāsa, the king of Kāshī. Pratardana thought himself an erudite scholar, because he studied the science of logic, *hetu vidyā*. Pratardana was clever. Wearing a red hat and a red silk shawl, he proudly debated in the king's court and triumphed over his opponents by revealing their fallacies in reasoning (*hetvābhāsha*).

Even though a prince, he had few friends, for he bewildered all with his sharp intellect. His heart was unmoved by the pure water of the Ganges and the sacred temples of Kāshī, for his mind was busy explaining riddles. Pratardana was a solver of puzzles.

Prince Pratardana wished to discover the source of logic, and so he decided to visit Indra, who is wholeness, the essence of all knowledge. However, because his heart was not pure, it was not easy for Pratardana to reach Indra. He had many obstacles to overcome.

But Pratardana persisted, and after a long journey and many hardships, he traveled to the eastern quarter of the sky, beyond the earth. There he approached *Svarga Loka,* the Realm of Light.

He climbed up a narrow mountain path, and as he reached the summit, wide valleys and rolling meadows suddenly stretched before him. Birds sang with joy, and rare blossoms shared their sweet fragrances. He heard Gandharvas playing enchanting music, and saw celestial beings enjoying many pleasures according to the fruits of their past actions.

He walked on to the city of Indra, called *Amarāvatī,* the abode of the immortals. In the city was the lovely Nandana grove, filled with dark blue butterflies flying to and fro in the sweet aroma of sandalwood. In the center of the garden stood the celebrated wish-fulfilling tree, Kalpa Vṛiksha. One could forget all his worries just by sitting in the shade of its crimson blossoms. Resting under the tree stood Kāmadhenu, the cow of plenty, who could fulfill any desire. It was said that the milk from her udder filled the ocean at the beginning of creation.

In a magnificent palace nearby, Pratardana met Indra, king of the Devas, who was drinking soma juice. Next to Indra stood his white horse, Uchchaiḥshravas, and his four-tusked elephant, Airāvata. Grasping his thunderbolt in his hand, Indra could create lightning with the glance of his eyes. Around his neck hung a long necklace of white pearls. It was a special necklace, for hidden inside each pearl nestled all the other pearls. Thus to see one pearl was to see them all.

Indra knew that Pratardana had made a great effort to reach him, and he said, "Pratardana, choose a boon for yourself!"

Pratardana thought of all the wonderful things he could wish for. But he said to Indra, "You yourself choose a boon for me."

"I am above you," Indra replied, "and you are below me. It is not right for me to choose for you. You must choose a boon for yourself."

"If I choose," persisted Pratardana, "then it will not be a boon."

Then Indra, who could speak only the truth, said, "Know the Self. This is the highest boon, which is most beneficial for you and for all mankind. Whoever knows the Self will not be injured. He will desire only the good, and so the color will never pale from his face."

Then Indra said,

"Wholeness is the breathing spirit. Wholeness is the Self. Know wholeness as life, as immortality, *amṛita*.

"Life is breath and breath is life. As long as breath is in the body, so long is there life. With the breathing spirit one obtains long life in this world. He who knows the Self as life, enjoys all the senses in this world and immortality in the next."

Pratardana did not understand. He asked,

"Does each sense have its own source, or do all the senses come from a single source?"

"If not a single source," replied Indra, "then how could the eyes see? How could the ears hear? How could the voice speak? How could the mind think? All these are unified at their source, which is life.

"When we speak, life speaks.
When we see, life sees.
When we hear, life hears.
When we think, life thinks."

Pratardana did not understand. He said,

"One can live without speech,
for there are the dumb.
One can live without sight,
for there are the blind.

"One can live without hearing,
for there are the deaf.
One can live without a good mind,
for there are the childish."

"But without breath one cannot live," said Indra. "And breath receives its life from consciousness.

"What is the breath of life?
Pure consciousness.
What is pure consciousness?
The breath of life.
This is the right view.
This is the right understanding."

"What is the relationship between consciousness and the senses?" asked Pratardana.

Indra said,

"When consciousness governs speech,
through speech we may speak all names.
When consciousness governs the eye,
through the eye we may see all forms.

"When consciousness governs the ear,
through the ear we may hear all sounds.
When consciousness governs the tongue,
through the tongue we may taste all flavors.

"When consciousness governs the feet,
through the feet we may make all movement.
When consciousness governs the mind,
through the mind we may think all thoughts."

"Can we live without consciousness?" asked Pratardana.

Indra replied,

"Without consciousness, indeed,
speech would not speak any name at all.
We would say, 'My mind was somewhere else.
I could not speak that name.'

"For indeed, without consciousness,
the eyes would not see any form at all.

We would say, 'My mind was somewhere else.
I could not see that form.'

"For indeed, without consciousness,
the ears would not hear any sound at all.
We would say, 'My mind was somewhere else.
I could not hear that sound.'

"For indeed, without consciousness,
the tongue would not taste any flavor at all.
We would say, 'My mind was somewhere else.
I could not taste that flavor.'

"For indeed, without consciousness,
the feet would not make any movement at all.
We would say, 'My mind was somewhere else.
I could not make that movement.'

"For indeed, without consciousness,
the mind would not think any thoughts at all.
We would say, 'My mind was somewhere else.
I could not think that thought.' "

Pratardana was still confused. He said, "To know what we speak, we should know speech. To know what we see, we should know the seen."

Indra replied,

"It is not speech we should know;
we should know the speaker.
It is not the seen we should know;
we should know the seer.

"It is not sound we should know;
we should know the hearer.
It is not taste we should know;
we should know the taster.

"It is not movement we should know;
we should know the mover.
It is not thoughts we should know;
we should know the thinker.

"Just as on a chariot, the spokes of the wheel
are fixed to the hub, so all aspects of intelligence
are fixed to the breathing spirit. This is the immortal—
the peaceful, the blissful, the undivided.

"One does not become great merely by good actions.
Knowing the Self, indeed, one rises up
and performs what is good.

"The protector of the world,
the sovereign of the world—
it is the Self. This you should know.
It is the Self. This you should know."

Then Pratardana lost his pride, because he found what he was searching for—himself. He became the king of Kāshī and governed with a fair hand, knowing that he and his subjects drew from the same breath of life. He saw life as wholeness, not as a puzzle to be solved.

Waist-deep in the sacred Ganges, Pratardana offered a cupped handful of water to the sun, as a small flotilla of marigolds passed. He smelled jasmine and sandalwood, and his eyes were wet with tears. He felt the beauty in life, and the beauty was bigger that his logic. For now his logic was rooted in wholeness, and what he perceived were ripples in that ocean of wholeness, which flows on forever, like the Ganges past the sacred temples in the city of Kāshī.

शिवं शान्तमद्वैतं
चतुर्थं मन्यन्ते
स आत्मा स विज्ञेयः

Shivaṃ shāntam advaitaṃ
chaturthaṃ manyante
sa Ātmā sa vigyeyaḥ

The peaceful, the blissful, the undivided
is thought to be the fourth;
that is the Self. That is to be known.
Nṛisiṃhottaratāpanīya Upanishad, 1

The Self is all this

CHAPTER NINETEEN

Nārada Visits Sanatkumāra, the Eternal Youth

From the Chhāndogya Upanishad

Once there was a wise scholar by the name of Nārada. Nārada was one of the seven Ṛishis of ancient India. He could recite from memory the four Vedas and many other Vedic texts. Nārada taught Vālmīki how to meditate, and Vālmīki later recorded the story of Rāma and Sītā, called the Rāmāyaṇa. Besides teaching Vālmīki, Nārada was famous for his discourse on right action, called the Nārada Smṛiti.

Nārada was also a great musician. When the night met with morning and the afternoon met with evening, he plucked delicate notes on his vīṇā, the seven-stringed instrument that resembles the sitār.

Every morning, Nārada sat on his couch, made of *khadira* wood, surrounded by students. He taught them the music of Gandharva Veda. He played the scale—*sa, re, ga, ma, pa, dha, ni, sa*—and then the students repeated it. Nārada played variations of the scale, and the students imitated the variations. He sang melodies, called *rāgas,* and the students sang rāgas. His fingers tapped rhythms, called *tālas,* and the students tapped tālas. Nārada taught students that the purpose of music is enchantment (*mohana*).

His students often brought him gifts of sweets, garlands, and woolen shawls. The wealthy among the villagers gave him cows. Thus Nārada was not only a famous and respected teacher, he was also wealthy.

However, Nārada knew in his heart that he was not completely content. For even though he was known throughout the land for his brilliant music, he desired lasting inner happiness.

"Who can teach me the knowledge of real happiness?" Nārada wondered.

After some time he decided to visit Sanatkumāra, who, it was said, was born of the mind of Brahmā, the Creator. Nārada knew that Sanatkumāra understood *Ātma Gyāna,* the knowledge of the Self, and through this knowledge he had conquered aging. His name meant "eternal youth," for due to his wisdom Sanatkumāra was blessed with everlasting youthfulness.

Nārada traveled to the heavenly world of *Jana Loka,* and there met Sanatkumāra. Even though Sanatkumāra had lived a long time, he appeared to be a young boy, with a resplendent face and bright eyes. Nārada greeted him humbly and said, "Dear Sanatkumāra, would you teach me Ātma Gyāna, the knowledge of the Self?"

Sanatkumāra spoke softly to Nārada. "First please tell me what you know," he said.

"I know by heart the Ṛik, Sāma, Yajuḥ, and Atharva Veda," said Nārada. "I know the Rāmāyaṇa and

Mahābhārata, as well as the eighteen Purāṇas. I know Vyākaraṇa (grammar), the Veda of Vedas. I know mathematics, logic, ethics, and politics. I know Nakshatra Vidyā, the science of the heavenly bodies. I know the art of music, Gandharva Veda, and the other fine arts. This, Sanatkumāra, is what I know."

Then Nārada added, "In knowing this, I know only the words. I am not a knower of the Self, *Ātmavit*. I have heard that 'Established in the Self, one overcomes sorrows and suffering' (*Tarati shokam Ātmavit*). Yet I am one who has sorrow. Help me to cross over to the other side."

"Yes, what you have learned is only the words," Sanatkumāra replied. "Those who know only words fulfill their desires in words only."

"Sanatkumāra, what goes beyond words?" Nārada asked.

"Speech (*vāk*) assuredly goes beyond words," Sanatkumāra said.

> "Without speech one would not know right from wrong,
> true from false, good from bad,
> or the pleasing from the unpleasing.
> Yet those who know only speech
> fulfill their desires in speech only."

"Sanatkumāra, what goes beyond speech?" Nārada asked.

"Mind (*manas*) assuredly goes beyond speech," Sanatkumāra said.

"As the closed fist holds two *āmalaka* fruits,
so the mind holds words and speech.
It is through the mind that one learns the Veda.
Through the mind one performs good actions.
Through the mind one obtains wealth.
Through the mind one knows this world
and what is beyond.
Yet those who know only the mind
fulfill their desires in mind only."

"Sanatkumāra, what goes beyond the mind?" Nārada asked.

"Intention (*saṃkalpa*) assuredly goes beyond the mind," Sanatkumāra said.

"Through intention one reflects and speaks.
Through intention heaven and earth were formed.
Through intention air and space were formed.
Through intention water and fire were formed.
Through intention one forms the world he desires,
and through intention becomes established in that world.
Yet those who know only intention
fulfill their desires in intention only."

"Sanatkumāra, what goes beyond intention?" Nārada asked.

"Meditation (*dhyāna*) assuredly goes beyond intention," Sanatkumāra said.

"The earth is meditating, as it were.
The heavens are meditating, as it were.
The waters are meditating, as it were.
The mountains are meditating, as it were.
God is meditating, as it were.
Those who meditate attain greatness
through their meditations.

"Through meditation one realizes the unbounded.
That which is unbounded is happy
There is no happiness in the small.
Only the unbounded is happy.
Be intent upon the unbounded.
Desire the unbounded.
Seek the unbounded.

"When one sees nothing other,
hears nothing other,
understands nothing other,
that person is unbounded.

"When one sees other than the unbounded,
that person is yet small.
The unbounded is all-pervading.
The unbounded is the immortal Self.
It is established in its own magnitude.
It is established in its own majesty.

"The unbounded is one. It becomes threefold,
fivefold, sevenfold, and ninefold.
Then again it becomes elevenfold,
and a hundred and eleven, and twenty thousand.
Sa ekadhā bhavati tridhā bhavati
panchadhā saptadhā navadhā chaiva
punash chaikādashaḥ smṛitaḥ
shataṃ cha dasha chaikash cha sahasrāṇi cha viṃshatiḥ

"It is above and below.
It is in front and behind.
It is to the north and to the south.
Indeed, it is all this.
Sa evādhastāt sa uparishtāt
sa pashchāt sa purastāt
sa dakshiṇataḥ sa uttarataḥ
sa evedaṃ sarvam

"I am above and below.
I am in front and behind.
I am to the north and to the south.
Indeed, I am all this.
Aham evādhastāt aham uparishtāt
ahaṃ pashchāt ahaṃ purastāt
ahaṃ dakshinato 'ham uttarataḥ
aham evedaṃ sarvam

"The Self is above and below.
The Self is in front and behind.
The Self is to the north and to the south.
Indeed, the Self is all this.
Ātmaivādhastād ātmoparishtāt
ātmā pashcād ātmā purastāt
ātmā dakshiṇata ātmottarataḥ
ātmaivedaṃ sarvam

"He who sees this, who thinks this,
who understands this—
he rests in the Self.
He delights in the Self.
He is united with the Self.
He has happiness in the Self.
He is his own ruler.
He fulfills his desires in every world.
Sa vā esha evaṃ pashyann evaṃ manvāna evaṃ vijānann
ātma-ratir
ātma-krīda
ātma-mithuna
ātmānandaḥ
sa svarād bhavati
tasya sarveshu lokeshu kāma-chāro bhavati

"He who realizes this does not see death,
nor illness, nor suffering.
He who realizes this, realizes everything
and obtains everything everywhere.

Na pashyo mṛityuṃ pashyati
na rogaṃ nota duḥkhatām
sarvaṃ ha pashyaḥ pashyati
sarvam āpnoti sarvashaḥ

"O Nārada, with this you will obtain the farther shore. With this you will become immortal."

Then Sanatkumāra gave Nārada a great gift—the freedom to roam at will anywhere on earth or in heaven. Nārada thanked him for this and for the knowledge of the Self, and began his journey home. As Nārada was traveling, he reflected upon what Sanatkumāra had said, "The Self is in front and behind. The Self is all that there is. He who realizes this realizes everything."

On his way home, he came upon Mount Kailāsa. At the summit, near a lake filled with lotus blossoms, sat Shiva and his wife, Pārvatī. Next to them were their two sons, Kārttikeya and Ganesh. Nārada greeted the divine family and offered Shiva a gift, a celestial mango.

Shiva took the wondrous mango, and told his two sons, "Whichever of you is the first to circle the entire world, his is this celestial fruit." With that, Kārttikeya mounted his peacock and, like a bolt of lightning, was gone. Ganesh stood for a moment and then slowly circled his parents, Shiva and Pārvatī. "You are the whole world," he said. "By circling you, I have have traversed the entire world."

With this, his parents were pleased and offered him the mango. Nārada observed the delightful encounter, and thought about what Sanatkumāra had told him, "He who realizes this, realizes everything and obtains everything everywhere." He knew that one could obtain everything in the entire world just by knowing the Self. To know the Self is to encompass the world.

And so the great scholar Nārada learned the knowledge of the Self, Ātma Gyāna, from the eternal youth, Sanatkumāra. Nārada continued to teach music to his students. Besides teaching them to play with perfect skill, he taught them to breathe life into their music. He taught them to draw from deep within, and thus to play the rhythm of the heavenly spheres, the music of the soul.

आत्मैवेदं सर्वम्

Ātmaivedaṃ sarvam

The Self is all this.

Chhāndogya Upanishad, 7.25.2

CHAPTER TWENTY

The Āchārya's Message on the Last Day of Study

From the Taittirīya Upanishad

White sand separated the vast blue sea from the green jungle, like a strand of pearls between blue sapphires and green emeralds. The jungle was thick with sandalwood, mahogany, cashew-nut, and teakwood trees. Coconut and palmetto palms lined the sand.

A cluster of small bamboo cottages rimmed the edge of the jungle. These bungalows served as both home and school for the teacher, the āchārya. Students lived and studied in the school, called a *guru-kula.* The cottages were surrounded by gardens of hibiscus flowers and soft grass. Peacocks and parrots played in the gardens, and monkeys scampered up and down the palms.

Every day, on the grass between the cottages, the students, called brahmachāris, studied with their teacher. They ate with their teacher. They sang rāgas and told stories with their teacher. The students helped with the daily chores—chopping wood, carrying water, and milking the cows.

For twelve years the āchārya taught the same students. He and his wife cared for these students as their own children. The teacher clothed them and looked after their good health. He taught them how to grow food and milk the cows. Above all, the āchārya gave them knowledge.

The āchārya taught them to recite the four Vedas—Ṛik, Sāma, Yajuḥ and Atharva. He taught them correct pronunciation and meter. He gave them the many rules of Sanskrit grammar, the Veda of Vedas. He showed them how to perform the various yagyas to bring good fortune and balance in nature. He instructed them in the knowledge of Jyotish, the science of prediction. He taught logic and Vedānta. He read them the great epics, the Rāmāyaṇa and Mahābhārata, and he told them stories from the Purāṇas. He taught them mathematics, science, logic, ethics, music, art, and drama.

He taught them to exercise and play games. He showed them how to be honest and fair. He demonstrated the various postures of yoga. Most important of all, he taught them the supreme knowledge of wholeness—Brahma Vidyā.

One day in spring the students gathered around the teacher in the pavilion between the bungalows. They chattered with excitement, because this was a special day. This was their last day in the guru-kula. After twelve years of study, the students were no longer children. They were grown up now, and would be leaving. The teacher knew that he would miss his students dearly.

The āchārya began the lesson in the usual way:

"May Mitra bring us tranquillity.
May Varuṇa bring us rest.
May Aryaman bring us quietude.
May Indra and Bṛihaspati bring us peace.
May Vishṇu of wide strides bring us contentment.
Shaṃ no mitraḥ
shaṃ varuṇaḥ
shaṃ no bhavatv aryamā
shaṃ na indro bṛihaspatiḥ
shaṃ no vishṇur uru-kramaḥ

"I give honor to Brahman.
I give honor to you, O Vāyu.
You are indeed the visible Brahman.
Indeed of you, the visible Brahman, will I speak.
Namo brahmaṇe
namas te vāyo
tvam eva pratyakshaṃ brahmāsi
tvām eva pratyakshaṃ brahma vadishyāmi

"I will speak of what is right.
I will speak of what is true.
May that satisfy me.
May that satisfy the teacher.
May I be satisfied.
May the teacher be satisfied."
Ṛitaṃ vadishyāmi
satyaṃ vadishyāmi
tan mām avatu

tad vaktāram avatu
avatu mām
avatu vaktāram

"Now, dear students, please sit," said the āchārya. He and his wife had decorated the seats with roses and orchids for this special day. The āchārya offered them milk, butter, and honey. The students recited special verses from the Vedas and then bathed in the ocean. This ceremony, called *Samāvartana,* meant that now the students had become fully enlivened with knowledge, and were ready to return to their families.

Now it was time for the āchārya to give them a parting message:

"Speak the truth.
Do your Dharma, your natural duty.
Meditate every day.
Read the Vedas every day.
After offering to your teacher,
do not loosen the tie of affection.

"Radiate truth.
Follow Dharma, Natural Law.
Think of the welfare of others.
Enjoy prosperity.
Always continue your studies,
and be filled with devotion to God.

"Honor your mother as God.
Honor your father as God.
Honor your teacher as God.
Honor your guest as God.
Mātṛi Devo bhava
pitṛi Devo bhava
āchārya Devo bhava
athiti Devo bhava

"Be pure in thought and action.
Do actions you know to be good.
Be kind to those who are older.
Give to others with faith.
Do not give without faith.
Give generously.
Give modestly.
Give with sympathy.

"If you doubt an action, do as the enlightened do.
Judge your actions by those of virtuous people.
If someone has spoken against you,
behave with kindness and love.

"Live your life in bliss.
When you are full of bliss,
you will know what is right to do.
You will know your Dharma.

"This is the knowledge.
This is the teaching.
This is the secret of the Veda.
This is the instruction.

"I have spoken what is right.
I have spoken what is true.
That has satisfied me.
That has satisfied the teacher.
I am satisfied.
The teacher is satisfied."
Ṛitam avādisham
satyam avādisham
tan mām āvīt
tad vaktāram āvīt
āvīn mām
āvīd vaktāram

Then the students presented gifts to their beloved teacher. They touched his feet and gave him woolen shawls, sweets, and garlands of orchids. They recited poems and sang songs.

"Dear Sir," one student said, "you are established in Brahman and learned in the Vedic scriptures. Thank you for teaching us with sweet speech and with conviction. "

Another student said, "We were like the blindfolded who could not find their way home without someone to remove the blindfolds. Thank you for removing our ignorance."

"We could not have learned without our beloved teacher," said still another student. "For even someone as intelligent and well-read as King Janaka needed a teacher like you."

"We owe you a great debt for all the wisdom you have given us," said another student. "How can we ever repay you?"

The āchārya smiled fondly and replied, "You can repay me by having students of your own, and teaching them wisdom."

After much celebration, the teacher and his wife said good-bye to their students. When everyone was gone, and all was quiet, the teacher walked alone by the sea. He thought about his dear students and about how many happy moments they had spent together over the past twelve years.

In his mind's eye, he could see each student's face. He could see their faith and trust in him. He could see each of them repeating after him like frogs croaking in the rainy season—*saha nāv avatu, saha nau bhunaktu*. . . . He could see the students touching his feet and offering their thanks for the knowledge they had received.

Just as students seek a good teacher, he knew that teachers also are happiest when they have ideal students. These students had not only promptly followed his wishes, but often anticipated them.

He thought of how proud he was of his students. How untroubled their minds were. How peaceful and tranquil they were. How they delighted in the Self.

He had often prayed for such students,

"May students come to me to learn the Veda.
May students come to me from all directions.
May students come to me with sincerity.
May students who are self-controlled come to me.
May students who are peaceful come to me.
May they be my wealth.
May they be my fame.
Ā māyantu brahmchāriṇaḥ svāhā
vi māyantu brahmchāriṇaḥ svāhā
pra māyantu brahmchāriṇaḥ svāhā
da māyantu brahmchāriṇaḥ svāhā
sha māyantu brahmchāriṇaḥ svāhā
yasho jane 'sāni svāhā
shreyān vasyaso 'sāni svāhā

"As water flows downwards,
as the months flow into the year,
may students come to me from all directions."
Yathāpaḥ pravatā yanti
yathā māsā aharjaram
evaṃ māṃ brahmachāriṇaḥ
dhātarāyantu sarvatas svāhā

He wandered for some time by the edge of the blue sea, lost in thought. And then suddenly he remembered, "Oh, I forgot!"

He turned and quickly hurried back to the school. There was much to be done. A new group of young students would arrive—tomorrow.

सह नाववतु
सह नौ भुनक्तु
सह वीर्यं करवावहै
तेजस्वि नावधीतमस्तु
मा विद्विषावहै

Saha nāv avatu
saha nau bhunaktu
saha vīryaṃ karavāvahai
tejasvi nāv adhītam astu
mā vidvishāvahai

Let us be together.
Let us eat together.
Let us be vital together.
Let us be radiating truth,
radiating the light of life.
Never shall we denounce anyone,
never entertain negativity.
Taittirīya Upanishad, Shānti Pātha

Glossary of Sanskrit Words—Pronunciation and Meaning

A

āchārya	(ah char´ yuh) teacher
Āditya	(ah di´ tyuh) the sun
Agni Jātavedas	(ug´ nee jah´ tuh vay´ dus) fire, one of the powers of nature
Agnihotra	(ug´ nee ho´ truh) a particular yagya
āhavanīya	(ah´ huh vuh nee´ yuh) the eastern fire used in yagyas
Airāvata	(ai rah´ vuh tuh) the white elephant of Indra
Aitareya	(ai´ tuh ray´ yuh) one of the ten principal Upanishads; also an Āraṇyaka; Mahīdāsa Aitareya learned the Vedas from his father, a Brāhmaṇa seer.
Ajātashatru	(uh jah´ tuh shuh´ troo) the famous king of Kāshī whose name means "whose enemies (*shatru*) are unborn (*ajāta*)."
ākāsha	(ah kah´ shuh) space
akshara samāmnāya	(uh´ kshuh ruh suh mam´ nah´ yuh) "recitation of letters," the Sanskrit alphabet
āmalaka	(ah´ muh luh kuh) small round fruit with a sour taste
Amarāvatī	(uh muh rah´ vuh tee) "the abode of the immortals," the city of Indra
amṛita	(um´ ri tuh) immortal, immortality
amṛitam	(um´ ri tum) nectar of immortality

ānanda (ah´ nun duh) bliss

anna (uhn´ nuh) matter, food

anudātta (uh´ noo dat´ tuh) "unraised," the low tone; marked in the Ṛik Saṃhitā by a horizontal bar below the syllable

anusvāra (uh noo swah´ ruh) a nasal sound, written ṃ

apāna (uh pah´ nuh) "downward breath," responsible for elimination

apara Brahman (uh´ puh ruh bruh´ mun) the lower Brahman, the manifest world

āpas (ah´ pus) water

Ara (uh´ ruh) the lake in Brahma Loka

Ārtabhāga (ar´ tuh bhah´ guh) one of the pandits in the debate of King Janaka

Aryaman (ar´ yuh mun) the sun, one of the powers of nature

āshram (ahsh´ rum) Vedic school (*āshrama* in Sanskrit)

Ashvala (ush´ vuh luh) King Janaka's pandit

Āshvalāyana (ash´ wuh lah´ yuh nuh) the man who visited Brahma Loka

Āshvalāyana Gṛihya Sūtra (ash´ wuh lah´ yuh nuh grih´ yuh soo´ truh) sūtras about yagyas by Āshvalāyana, included in the Kalpa section of the Vedic Literature

Āshvalāyana Shrauta Sūtra (ash´ wuh lah´ yuh nuh shrow´ tuh soo truh) sūtras about yagyas by Āshvalāyana, included in the Kalpa section of the Vedic Literature

Ashvapati (ush´ wuh puh´ tee) "lord of horses," the King of Kekaya who taught the learned scholars about Ātmā. He was the father of Kaikeyī.

asura	(uh soo´ ruh) negative power of nature
Atharva	(uh thar´ vuh) one of the four Vedas
Ātmā	(aht´ mah) the Self, pure consciousness, pure awareness
Ātma Gyāna	(aht´ muh gyah´ nuh) knowledge of the Self
Ātma Vidyā	(aht´ muh vi´ dyah) knowledge of the Self
Ātmavit	(aht´ muh vit´) knower of the Self
Ayodhyā	(uh yo´ dhyah) city ruled by King Dasharatha

B

Badrīnāth	(buh´ dree nath) shrine in the Himālayas
Bālāki	(bah´ lah kee) the proud teacher
bhāra	(bhah´ ruh) a particular weight
Bhārgava	(bhar´ guh vuh) one of the six friends who learned from Pippalāda
Bhṛigu	bhri´ goo) the son of Varuṇa
Bhujyu	(bhoo´ jyoo) one of the pandits in the debate of King Janaka
bhūma	(bhoo´ muh) the unbounded
Brahmā	(bruh´ mah) the power of nature responsible for creation, the Creator
brahma-muhūrta	(bruh´ muh moo hoor´ tuh) the time of Lord Brahmā, a period of 48 minutes before dawn
Brahma Loka	(bruh´ muh lo´ kuh) the realm of Brahmā, the Creator
Brahma Vidyā	(bruh´ muh vi´ dyah) knowledge of the wholeness of life
brahmacharya	(bruh´ muh char´ yuh) student life
brahmachārī	(bruh´ muh chah´ ree) student, life-long student
Brahman	(bruh´ mun) wholeness, totality
brahmasthān	(bruh´ muh sthan´) "place of Brahman," the center of a kingdom, city, or home.

Brahmin	(brah´ min) teacher and scholar (*brāhmaṇa* in Sanskrit)
Bṛihadāraṇyaka	(bri´ hud ah´ run yuh kuh) one of the ten principal Upanishads
Bṛihaspati	(bri hus´ puh tee) one of the powers of nature
Budha	(boo´ dhuh) the planet Mercury
Budila	(boo´ dee luh) one of the scholars who learned about the Self from King Ashvapati

C

chakora	(chuh ko´ ruh) a green partridge that lives on moonbeams
chakshu	(chu´ kshoo) sight
Chandramā	(chun´ druh mah) the moon
Chhāndogya	(çhan do´ gyuh) one of the ten principal Upanishads
chid-ānanda	(chi dah´ nun duh) bliss consciousness

D

dakshiṇa	(duh´ kshee nuh) the southern fire used in yagyas; also called *anvāhāryapachana*
dakshiṇā	(duh´ kshee nah) the gifts given to the Brahmins who perform a yagya
damaru	(duh´ muh roo) the drum held by Lord Shiva
Dashāraṇya	(duh shah´ run yuh) forest where Pippalāda meditated for a long time
Dasharatha	(duh´ shuh ruh´ thuh) King of Ayodhyā and father of Rāma
Deva	(day´ vuh) positive power of nature
devanāgarī	(day´ vuh nah´ guh ree) "city (*nāgarī*) of immortals (*deva*)," also the name of the script used in the city of the immortals.

Dhanvantari (dhun vun´ tuh ree) "moving in a curve," the physician of the gods

Dharma (dhar´ muh) Natural Law, natural duty

dhyāna (dhyah´ nuh) meditation

dhyānya (dhyan´ yuh) corn, grain

diva (dee´ vuh) the sky

Divodāsa (dee´ vo dah´ suh) a famous king of Kāshī; father of Pratardana

droṇa (dro´ nuh) a wooden vessel, a certain measure

G

Gandharva (gun dhar´ vuh) celestial musician or singer

Gandharva Veda (gun dhar´ vuh vay´ duh) the section of Vedic Literature concerned with music

Gaṇesh (guh nesh´) son of Shiva and Pārvatī who had the head of an elephant; the force of nature who removes obstacles and brings good fortune (*Gaṇesha* in Sanskrit)

Gangā (gun´ gah) the Ganges River (known by many names, such as the River of Heaven, the Granter of Wishes, the River of Life, and the Stream of Nectar)

Gārgī (gar´ gee) the famous woman teacher in the debate of King Janaka

Gārgya (gar´ gyuh) one of the six friends who learned from Pippalāda

gārhapatya (gar´ huh puh´ tyuh) the western fire used in yagyas

gārhasthya (gar hus´ thyuh) householder life

Gautama (gow´ tuh muh) the teacher of Satyakāma

gāyatrī (gah´ yuh tree) the first of the seven traditional meters, composed of three lines, each with eight syllables ghāt — (ghat) area with stone steps leading down into a river

gotra — (go´ truh) Vedic family

guru-kula — (goo´ roo koo´ luh) home and school of the teacher

H

hetu vidyā — (hay´ too vi´ dyah) "knowledge of causes," the science of reasoning, logic

hetvābhāsha — (hay´ tvah bhah´ shuh) fallacy in reasoning

Himālaya — (hi mah´ luh yuh) "abode (*ālaya*) of snow (*hima*)," the great snow-capped mountain range of northern India

Hiraṇyagarbha — (hi run´ yuh gar´ bhuh) the golden (*hiraṇya*) womb (*garbha*) of creation

I

Indra — (in´ druh) the leader of the Devas

Indradyumna — (in´ druh dyoom´ nuh) one of the scholars who learned about the Self from King Ashvapati

Īsha — (ee´ shuh) "Lord," one of the ten principal Upanishads

Itihāsa — (i´ tee hah´ suh) "so it was," history; the two great epics of India, the Rāmāyaṇa and Mahābhārata

J

Jabālā — (juh bah´ lah) the mother of Satyakāma

jambu — (jum´ boo) the immense rose-apple tree (*eugenia jambolana*) on Mount Meru, whose fruits were the size of elephants (The English word "jumbo" comes from the

	name of a large elephant exhibited by P. T. Barnum.)
Jana	(juh´ nuh) one of the scholars who learned about the Self from King Ashvapati
Jana Loka	(juh´ nuh lo´ kuh) "the world of men," home of Sanatkumāra
Janaka	(juh´ nuh kuh) the famous king of Videha
Jānashruti	(jah´ nuh shroo´ tee) the king whose name means "celebrated (*shruti*) among the people (*jāna*)."
jyoti	(jyo´ tee) light
Jyotish	(jyo´ tish) the Vedic science of prediction (*Jyotisha* in Sanskrit)

K

Kabandhī	(kuh bun´ dhee) one of the six friends who learned from Pippalāda
Kaikeyī	(kai kay´ yee) daughter of King Ashvapati, from Kekaya
Kailāsa	(kai lah´ suh) mountain in the Himālayas which is the home of Shiva and Pārvatī
Kaivalya	(kai vul´ yuh) "singularity," unity, name of an Upanishad
kāla	(kah´ luh) time
kalash	(kuh´ lush) the golden vessel containing amṛitam (*kalasha* in Sanskrit)
Kaḷpa Vṛiksha	(kul´ puh vri´ kshuh) the wish-fulfilling tree (*erythrina indica*) in Svarga Loka, also called Pārijāta
Kāmadhenu	(kah´ muh dhay´ noo) the cow of plenty, the wish-fulfilling cow, also called Surabhi
Kamala	(kuh´ muh luh) the father of Upakosala
karma	(kar´ muh) action, the result of action

Kārttikeya	(kar´ ti kay´ yuh) son of Shiva and Pārvatī; also called Skanda or Subramaṇya
Kāshī	(kah´ shee) the sacred city on the Ganges. Kāshī is the ancient name of Vārāṇasī, also called Banāras in modern times.
Katha	(kuh´ thuh) one of the ten principal Upanishads
Kausalya	(kow sul´ yuh) one of the six friends who learned from Pippalāda
Kaushītaki	(kow´ shee tuh kee) one of the Upanishads
Kedārnāth	(kay´ dar nath) shrine in the Himālayas
Kena	(kay´ nuh) "by whom," one of the ten principal Upanishads
khadira	(khuh´ dee ruh) a hardwood (*acacia catechu)*
Kshatriya	(kshuh´ tree yuh) warrior or administrator
Kshiprā	(kshi´ prah) one of the rivers in central India
Kumbha Melā	(koom´ bhuh may´ lah) the large festival held every twelve years in Allahabad
Kuru	(koo´ roo) one of the ancient kingdoms of India; also an ancient family of India

L

lakshaṇa	(luk´ shuh nuh) auspicious mark

M

Maghavan	(muh´ ghuh vun) "bountiful," a name for Indra
Mahābhārata	(muh hah´ bhah´ ruh tuh) "Great India," one of the two great epics of India
mahājanapada	(muh hah´ juh´ nuh puh´ duh) great kingdom. There were said to be sixteen in ancient India.
Mahārāja	(muh hah rah´ juh) great king

mahāshāla	(muh hah´ shah´ luh) great householder
mahāshrotriya	(muh hah´ shro´ tree yuh) great scholar
Mahīdāsa Aitareya	(muh hee´ dah´ suh ai´ tuh ray´ yuh) the boy who learned about creation in the Aitareya Upanishad. He is said to have lived 116 years.
Maitreyī	(mai tray´ ee) the beloved wife of Yāgyavalkya
manas	(muh´ nus) mind
Manovatī	(muh no´ vuh tee) the city of Brahmā in the center of Brahma Loka
Meru	(may´ roo) the mountain on which Brahma Loka is located
Mithilā	(mee´ thee lah) capital city of Videha, where King Janaka ruled
Mitra	(mi´ truh) "friendship," one of the powers of nature
mohana	(mo´ huh nuh) enchantment
Mṛityu	(mri´ tyoo) death, one of the powers of nature
Muṇḍaka	(moon´ duh kuh) one of the ten principal Upanishads

N

Nachiketas	(nuh chi kay´ tus) the son of Vājashravas (sometimes written Nachiketa)
Nakshatra Vidyā	(nuk shuh´ truh vi´ dyah) the science of the heavenly bodies, Jyotish
nāma	(nah´ muh) name
namaste	(nuh´ mus tay') "I bow down to you," a phrase for greeting someone
Nandana	(nun´ duh nuh) the beautiful garden in the abode of Indra

Nārada	(nah´ ruh duh) the wise sage who learned about happiness from Sanatkumāra
Nārada Smṛiti	(nah´ ruh duh smri´ tee) text on Dharma by Nārada
Natarāja	(nuh´ tuh rah´ juh) the lord of dance, Shiva
Nṛisiṃhottaratāpanīya	(nri sing hot´ tuh ruh tah´ puh nee yuh) one of the Upanishads
nyagrodha	(nyuh gro´ dhuh) "growing downward," the banyan tree (*ficus indica*), whose branches send down roots which form new stems

P

pala	(puh´ luh) a particular measure
Panchāla	(pun chah´ luh) one of the kingdoms of ancient India
pandit	(pun´ dit) learned scholar, one who performs yagyas and chants the Vedas (*paṇḍita* in Sanskrit)
Pāṇini	(pah´ nee nee) sage who heard the Sanskrit alphabet and composed the fourteen Shiva Sūtras, which begin his treatise on Sanskrit
para Brahman	(puh´ ruh bruh´ mun) the higher Brahman, pure consciousness
Paramātma	(puh´ rum aht´ muh) the transcendental Self, the highest Self
Parameshvara	(puh´ ruh maysh´ wuh ruh) highest lord
Pārvatī	(par´ vuh tee) wife of Shiva
pippala	(pip´ puh luh) the fig tree (*ficus religiosa*), and the sweet berry from the tree; also called the *ashvattha* tree
Pippalāda	(pip puh lah´ duh) the teacher who answered questions from six students. He was said to be fond of the pippala berry.

Prāchīnashāla (prah´ chee nuh shah´ luh) one of the scholars who learned about the Self from King Ashvapati

Prajāpati (pruh jah´ puh tee) "lord of creation," the protector of life, one of the powers of nature

prāṇa (prah´ nuh) vital breath, the breath of life

Prashna (prush´ nuh) "question," one of the ten principal Upanishads

Pratardana (pruh´ tur duh´ nuh) the son of King Divodāsa who was the prince of Kāshī and visited Indra

Prātishākhya (prah´ tee shah´ khyuh) various aspects of the Vedic Literature, each belonging to one of the four Vedas

pṛithivī (pri´ thi vee) earth

Purāṇa (poo rah´ nuh) "ancient." There are eighteen Purāṇas in the Vedic Literature.

Purusha (poo´ roo shuh) universal Being, the unbounded Self

Pūshan (poo´ shun) the sun, one of the powers of nature

R

rāga (rah´ guh) melody, song

Raikva (rai´ kvuh) the enlightened cart driver

rājā (rah´ jah) king

rājaputra (rah´ juh poo´ truh) "king's son," prince

Rāma (rah´ muh) hero of the Rāmāyaṇa, son of King Dasharatha

Rāmāyaṇa (rah´ mah´ yuh nuh) "the story of Rāma," one of the two great epics of India

Ṛik (rik) the first of the four Vedas, Ṛik Saṃhitā

Ṛishi (ri´ shee) seer, sage

S

Sāma — (sah´ muh) "evenness," one of the four Vedas

samāna — (suh mah´ nuh) the even breath, responsible for digestion

Sāmashravas — (sah´ muh shruh´ vus) the young pupil and servant of Yāgyavalkya

Samāvartana — (suh mah´ var´ tuh nuh) the bathing ceremony at the end of one's studies

Saṃhitā — (sung´ hee tah) "unity, collectedness," the collected state of consciousness, pure consciousness; a collection of verses

saṃkalpa — (sung kul´ puh) intention, declared publicly before the beginning of a yagya

saṃskṛita — (sun´ skri tuh) "perfected," Sanskrit, the language of nature

Sanatkumāra — (suh´ nut koo mah´ ruh) "eternal (*sanat*) youth (*kunāra*)," the teacher of Nārada

sannyāsa — (sun nyah´ suh) retired life

Saraṣvatī — (suh´ ruh swuh´ tee) the wife of Brahmā

sarva-kāma-siddhi — (sar´ vuh kah´ muh sid´ dhee) "all (*sarva*) desires (*kāma*) accomplished (*siddhi*)," the boon given to Pippalāda that granted him anything he wished.

Satya Loka — (suh´ tyuh lo´ kuh) the realm of truth, another name for Brahma Loka

Satyakāma — (suh´ tyuh kah´ muh) the young boy whose name means "seeker (*kāma*) of truth (*satya*)"; the teacher of Upakosala. The Satyakāma who went with five friends to study with Pippalāda was a different Satyakāma, Satyakāma Shaibya.

Satyayagya	(suh´ tyuh yuh´ gyuh) one of the scholars who learned about the Self from King Ashvapati
shākhā	(shah´ khah) "branch," a version, or recension, of a Vedic text; sometimes applies to the school or branch that maintains that particular text
Shiva	(shee´ vuh) the power of nature responsible for evolution
Shiva Sūtra	(shee´ vuh soo´ truh) fourteen short aphorisms which contain the Sanskrit alphabet, heard by Pāṇini; also a later text on Yoga
shrotra	(shro´ truh) hearing
Shūdra	(shoo´ druh) laborer or sweeper
Shukra	(shoo´ kruh) "the pure," the planet Venus
Shvetaketu	(shvay´ tuh kay´ too) the son of Uddālaka Āruṇi
Shvetāshvatara	(shvay tash´ vuh tuh ruh) the Ṛishi who becomes a teacher, the name of an Upanishad in which he is the teacher
siddha	(sid´ dhuh) perfected being
Sītā	(see´ tah) the daughter of King Janaka, the wife of Rāma
Smṛiti	(smri´ tee) "memory," section of the Vedic Literature which teaches right action
soma	(so´ muh) the ambrosia of immortality, the juice used in performing a yagya; the moon
Sukesha	(soo kay´ shuh) one of the six friends who learned from Pippalāda
Sūrya	(soor´ yuh) the sun
sushupti	(soo shoop´ tee) deep sleep

sutejas — (soo tay´ jus) "auspicious (*su*) light (*tejas*)," bright light, splendor

sūtra — (soo´ truh) "thread," short aphorism, short phrase

svāhā — (svah´ hah') Hail! (said while making offerings to the yagya fire)

svapna — (swup´ nuh) dreaming state

Svarga Loka — (swar´ guh lo´ kuh) the realm of light, the abode of Indra

svarita — (swuh´ ree tuh) the high tone; marked in the Ṛik Saṃhitā by a vertical bar above the syllable

T

Taittirīya — (tait´ ti ree´ yuh) one of the ten principal Upanishads

tāla — (tah´ luh) rhythm

tapas — (tuh´ pus) "inner glow," meditation

Tārkshya — (tark´ shyuh) one of the powers of nature

tejas — (tay´ jus) light, spiritual brilliance

tīrtha — (teer´ thuh) holy shrine, usually near a body of water such as a river

turīya — (too ree´ yuh) "fourth," pure consciousness, pure awareness (the fourth state of consciousness)

U

Uchchaiḥshravas — (uch´ chay shruh´ vus) Indra's white horse

udāna — (oo dah´ nuh) the upward breath

udātta — (oo dat´ tuh) "raised," the middle tone; unmarked in the Ṛik Saṃhitā

Uddālaka Āruṇi — (oo dah´ luh kuh ah´ roo nee) scholar who lived in Panchāla, father of Shvetaketu udumbara (oo doom´ buh ruh) a tree (*ficus glomerata*) with fruit as sweet as honey

Umā (oo´ mah) the daughter of the Himālayas

Upakosala (oo´ puh ko´ suh luh) the student of Satyakāma who learned about Brahman from the yagya fires

Upanayana (oo´ puh nuh´ yuh nuh) the initiation ceremony, qualifying a person to study the Vedic Literature and practice meditation

Upanishad (oo puh´ ni shud) "to sit down near," an aspect of Vedic Literature concerned with the nature of pure consciousness.

V

vaishvānara (vaish´ vah´ nuh ruh) universal

Vaishya (vai´ shyuh) merchant or farmer

Vājashravas (vah´ jush ruh´ vus) the father of Nachiketas

vāk (vahk) speech

Vālmīki (val mee´ kee) composer of the Rāmāyaṇa

vānaprasthya (vah´ nuh prus´ thyuh) forest dweller life

Vārāṇasī (vah´ rah nuh see) the modern name of Kāshī

Varuṇa (vuh´ roo nuh) the father of Bhṛigu; also water, one of the powers of nature

Vāyu Mātarishvan (vah´ yoo mah´ tuh rish´ vun) wind or air, one of the powers of nature

Veda (vay´ duh) pure knowledge, the fundamental structures of Natural Law at the basis of the universe. "The Vedas" usually refers to the four Vedas: Ṛik, Sāma, Yajuḥ, and Atharva. "The Vedas" can also refer to the Vedic Literature as a whole.

Veda Bhūmi (vay´ duh bhoo´ mee) the land of knowledge, India

Vedānga (vay dang´ guh) section of the Vedic Literature. There are six Vedāngas: Shikshā, Kalpa, Vyākaraṇa, Nirukta, Chhandas, and Jyotish.

Vedānta (vay dan´ tuh) "culmination of the Veda," an aspect of the Vedic Literature

Vedavāṇī (vay´ duh vah´ nee) the language of nature, the language of the Veda; sometimes called Devavāṇī, the speech of the Devas

vedi (vay´ dee) the elevated ground in the center of the yagya hall

Videha (vi day´ huh) the kingdom in ancient India ruled by King Janaka

Vidyut (vi´ dyoot) lightning

vigyāna (vi gyah´ nuh) intelligence

vīṇā (vee´ nah) seven-stringed instrument that resembles the sitār

Virochana (vi ro´ chuh nuh) one of the asuras

Vishṇu (vish´ noo) the force of nature that maintains creation

Vishvanātha (vish´ vuh nah´ thuh) the most famous temple in Vārāṇasī

Vyākaraṇa (vyah´ kuh ruh nuh) grammar, one of the six Vedāngas

vyāna (vyah´ nuh) the moving breath, responsible for circulation

Y

yagya (yuh´ gyuh) performance that creates balance in nature

yagya-shālā	(yuh´ gyuh shah´ lah) the hall constructed for the yagya
Yāgyavalkya	(yah´ gyuh vul´ kyuh) the famous teacher who wins the debate of King Janaka and who later instructed King Janaka
yajamāna	(yuh´ juh mah´ nuh) patron of the yagya
Yajuḥ	(yuh´ jooh) one of the four Vedas
Yama	(yuh´ muh) the sun; also the administrator of death and immortality
Yoga	(yo´ guh) union, the settled mind; also, the various practices for settling the mind, such as yoga āsanas and meditation
yogī	(yo´ gee) one who has attained Yoga
yojana	(yo´ juh nuh) a measurment of distance